Fluid Mechanics

College Work Outs for degree and professional students

Dynamics
Electric Circuits
Electromagnetic Fields
Electronics
Elements of Banking
Engineering Materials
Engineering Thermodynamics
Fluid Mechanics
Heat and Thermodynamics

Mathematics for Economists
Mechanics
Molecular Genetics
Numerical Analysis
Operational Research
Organic Chemistry
Physical Chemistry
Structural Mechanics
Waves and Optics

Fluid Mechanics

G. Boxer

MACMILLAN

First published 1988 by
THE MACMILLAN PRESS LTD
Houndmills, Basingstoke, Hampshire RG21 2XS
and London
Companies and representatives
throughout the world

ISBN 0–333–45122–8\

13 12 11 10   9   8   7   6   5
03 02 01 00 99 98 97 96 95

Printed in Malaysia

Contents

Preface   vii
Nomenclature   viii
Introduction   xi
       General Approach to Problem Solving   xi
       Some General Definitions in Fluid Mechanics   xiv

1   Fundamental Properties of a Real Fluid   1
       1.1   Introduction   1
       1.2   Worked Examples   4

2   Hydrostatic Pressure   8
       2.1   Introduction   8
       2.2   Manometry   9
       2.3   Hydrostatic Forces on Submerged Surfaces   12
       2.4   Worked Examples   16

3   The Steady, One-dimensional Flow of an Ideal Incompressible Fluid   31
       3.1   Introduction   31
       3.2   One-dimensional Equation of Continuity of Flow   32
       3.3   Euler's Equation for an Ideal Fluid in One-dimensional Flow   32
       3.4   Relationship Between Bernoulli's Equation and the Steady-flow Energy Equation   33
       3.5   Worked Examples   34

4   Flow Measurement   41
       4.1   Introduction   41
       4.2   Venturimeter   41
       4.3   Sharp-edged Orifice   43
       4.4   The Flow Nozzle   45
       4.5   The Pitot Tube   46
       4.6   The Pitotstatic Tube   46
       4.7   Worked Examples   47

5   The Momentum Equation for the Steady Flow of an Inviscid Incompressible Fluid   55
       5.1   Introduction   55
       5.2   Applications   57
       5.3   Worked Examples   62

6   Flow in Pipes of Viscous Incompressible Fluids   76
       6.1   Introduction   76
       6.2   Flow Between Two Reservoirs   78
       6.3   Power Transmission Through Pipelines   82

	6.4 Nozzle at Pipe Outlet	83
	6.5 Friction Factor	85
	6.6 Worked Examples	86
7	**Dimensional Analysis and Dynamical Similarity**	**99**
	7.1 Introduction	99
	7.2 Use of Dimensionless Groups	99
	7.3 Model Testing	100
	7.4 Buckingham's Pi Method	101
	7.5 Worked Examples	102
8	**Open-channel Flow**	**110**
	8.1 Introduction	110
	8.2 Manning Formula	111
	8.3 Flow-measuring Devices for Open-channel Flow — Notches and Weirs	111
	8.4 Worked Examples	114
9	**Steady Laminar Incompressible Viscous Flow**	**122**
	9.1 Incompressible Flow of Fluid Between Parallel Plates	122
	9.2 Shear Stress in a Circular Pipe with Steady Flow	123
	9.3 Laminar Flow Through a Circular Pipe	124
	9.4 Laminar Flow Between Stationary Parallel Pipes	126
	9.5 Dashpot	129
	9.6 Worked Examples	130
10	**The Steady, One-dimensional Flow of an Ideal Compressible Fluid**	**140**
	10.1 Introduction	140
	10.2 Stagnation Properties	141
	10.3 Acoustic Velocity and Mach Number	142
	10.4 Critical Pressure Ratio in a Nozzle	147
	10.5 Nozzle Efficiency	147
	10.6 Worked Examples	148
	Note on Specimen Examination Papers	**161**
11	**Examination Paper Number 1**	**162**
	11.1 Questions	162
	11.2 Suggested Solutions	164
12	**Examination Paper Number 2**	**172**
	12.1 Questions	172
	12.2 Suggested Solutions	174
	Recommended Reading	**182**
	Index	**183**

Preface

This is a revision tutorial book like its companion volumes in the Work Out Series. It is intended to be used in the period between the end of lectures in the early stages of a degree course and the start of sessional examinations. In certain cases a fairly full summary of fundamental lecture material is incorporated as students will want to compare the nomenclature in use here with that in their own notes or textbook.

A knowledge of elementary differential and integral calculus on the part of the student is taken for granted and the solutions are effected in the Système International since this is now in common use throughout the United Kingdom and on the European continent.

The fundamentals of fluid mechanics and engineering thermodynamics bear a close relationship to each other and the principles of conservation of mass, energy and momentum apply in each subject in very similar fashion. Indeed, attempts have been made elsewhere to combine the two subjects at first-year level to demonstrate this essential unity, but without real success. Where the two subjects are inextricably linked (e.g. in one-dimensional steady-flow gas dynamics – Chapter 10 in this book) the commonality is clearly depicted. Indeed, it is not until Chapter 10 that temperature becomes a significant variable and thermodynamics becomes an essential part of the analysis.

I have attempted wherever possible to use a nomenclature common to both thermodynamics and fluid mechanics, since they are companion volumes in this series and the two go hand in hand in any degree course in mechanical engineering.

The subject matter chosen should cover all that is likely to be included in the early stages of a degree course and the book is completed by the full solution of two sessional examination papers (as in *Work Out Engineering Thermodynamics*), so that students are forced to make their own decision concerning the nature of the question and the particular fundamentals involved without the prior benefit of chapter headings.

I have tried in the Introduction to stress the vital importance of a logical approach to all problems and the correspondingly vital need to maintain a correct dimensional balance in all reasoning. This cannot be too strongly emphasised. A great many difficulties can be resolved if the student forms the right habits from the word go.

I must pay tribute to two former colleagues of mine at the University of Aston for permission to include a considerable proportion of the material used in this book, namely Mr W. H. Kinsman and Mr J. C. Scouller.

The tutorial work presented herein has been used over a long period of time at Aston and is fully tested in the classroom. However, if any errors are discovered I will be very grateful if they are notified to me.

1988   G. B.

Nomenclature

A	area
A	location
a	dimension
a	atmospheric, acoustic or air (subscript)
B	dimension
B	location
b	dimension
C	centre line
C	Chezy constant or coefficient
	$(C_D/\sqrt{1 - (A_0/A_1)^2}$, where C_D = Discharge coefficient)
c	specific heat capacity
c_p	specific heat capacity at constant pressure
D	dimension
d	dimension
d	differential operator or location
D	location or discharge
E	internal energy
e	specific internal energy (= E/m)
e	location or equivalent (subscript)
F	force
f	friction factor
f	friction or (as a subscript) liquid state
G	centre of gravity
g	gravitational acceleration
g	saturated vapour state (as a subscript)
H	horizontal (subscript)
H	enthalpy or head
h	specific enthalpy (= H/m) or head or depth in a fluid
I	second moment of area
I_0	second moment of area about the centroid
i	inlet (to a pump)
i	slope of a channel
K	bulk modulus of elasticity
K	Kelvin (unit of temperature on the absolute scale)
L	length
l	length
M	fundamental dimension for mass
m	manometric or mean (subscript)
m	mass
m_v	relative molar mass
$\dot{m}$	mass flow rate (= dm/dt)

N	rotational speed
n	polytropic index
O	orifice
o	oil or conditions at solid boundary (subscript)
P	perimeter
p	pressure
P	pump, piston or plunger
Q	heat transfer
$\dot{Q}$	heat transfer rate $(= dQ/dt)$
q	specific heat transfer $(= Q/m)$
R	Rankine unit of absolute temperature
R	radius or specific gas constant
R_0	universal gas constant
r	radius
S	entropy or specific gravity
s	system
s	specific entropy $(= S/m)$ or distance
t	time or thickness (of an oil film)
t	throat
T	temperature
T	turbine (subscript)
u	velocity
V	volume
$\dot{V}$	volume flow rate $(= dV/dt)$
v	specific volume $(= V/m)$
V	vertical (subscript)
W	work transfer or weight
$\dot{W}$	rate of work transfer (i.e. power) $(= dW/dt)$
w	water (subscript)
w	specific work transfer $(= W/m)$
x	principal coordinate, distance or quality (dryness) of vapour
y	principal coordinate or dimension
Y	dimension
z	datum height (as in potential energy per unit mass gz)

α	angle
γ	isentropic index
δ	increment in (e.g. δx)
ϵ	excrescence height (average) in rough pipes
η	efficiency
θ	angle
μ	coefficient of absolute viscosity
ν	kinematic viscosity $(= \mu/\rho)$
π	as in area of circle (i.e. πr^2)
ρ	density
τ	shear stress
ω	angular velocity
Δ	increment in (e.g. Δx)
σ	surface tension
ϕ	function of

(Re)	Reynolds number $(= \rho u L/\mu)$
(Ma)	Mach number $(= u/u_a)$
(Fr)	Froude number $(= u^2/gL)$

head coefficient $= (g\,\Delta z)/u^2$
drag coefficient $= F_{\mathrm{D}}/(0.5\rho u^2 L^2)$
lift coefficient $= F_{\mathrm{L}}/(0.5\rho u^2 L^2)$

Other symbols

∂	partial differential operator
$\propto$	proportional to
$>$	greater than
$<$	less than
$\hat{\ }$	maximum value (e.g. $\hat{W}$)
∞	infinity
Σ	sum of
$\equiv$	equivalent to

Introduction

General Approach to Problem Solving

Fluid mechanics deals with the study of fluids (liquids and gases), both at rest (statics) and when flowing (fluid dynamics). The need to study fluid mechanics is paramount for the engineer and the applications are numerous and very varied. The flow of oil in a pipeline, of blood through a human body, of air round a propeller are but three examples of fluid dynamics. The forces on a masonry dam, the study of manometry and the use of hydraulic pressure to bend pipes are three obvious examples of fluid statics.

Very often the engineering problem involves the principles embodied in both fluid mechanics and thermodynamics and the trained engineer has learned to handle both simultaneously. However, having read and heard the arguments for a combined treatment of the two subjects at first-year level, I feel (as already stated in the Preface) that there is probably more to be gained by separate treatment at first, with appropriate cross-reference, leaving the dovetailing of the subjects to the more mature student. One obvious cross-reference is the link between the steady-flow energy equation and Bernoulli's equation which is demonstrated later in this book.

Like thermodynamics, the subject of fluid dynamics is often, misguidedly, subdivided in the student's mind into a series of discrete packets of theory, e.g. flow through venturimeters, manometry, and water meters, each containing its own individual formulae. In thermodynamics a comparably erroneous approach would similarly segregate piston engines, gas turbines, steam engines, and so on.

The student who wishes to master both subjects must learn to recognise that the same fundamental concepts are used repeatedly in both subjects, and that when these are absorbed the subsequent development of each is manifestly easier.

I have tried to emphasise this by postulating certain basic questions which should be asked every time a problem is attempted. I do not maintain that the questions are relevant to every problem encountered in engineering, since the range is so wide (there are obvious exceptions such as in manometry where they are of a limited assistance), but very often they give a clear guideline to a possible solution.

The first two questions are interrelated, but are given separately in the hope that students will be helped to a quicker understanding of the nature of the problem than they would be if the two questions were combined into a rather longer and more complex one.

(a) What Kind of *Process*?

In the subject of fluid mechanics an important distinction is between steady flow of a fluid (flow in which it may be assumed that properties across the cross-section of the fluid are constant with time, and mass flow rate is fixed), unsteady flow (mass flow rate a variable), and fluid statics (as in manometry).

(b) What Kind of *Fluid*?

In fluid mechanics the important distinction is made between incompressible (i.e. constant-density) fluids and compressible (variable-density) fluids. There will be occasions when fluids which are normally regarded as compressible, e.g. gases, are taken as incompressible since density changes are small (e.g. air flow round a lorry).

Furthermore, in the case of water-hammer, the liquid must be considered as compressible (whereas water is normally regarded as incompressible, especially in the problems in this elementary book) since the flow is characterised by the propagation of waves of finite amplitude with finite velocity.

(c) Have You Drawn a *Sketch* of Events?

Engineers work with pictures because they afford a valuable insight into the problem under examination. In thermodynamics the obvious picture is a field of state (e.g. pressure versus specific volume) and this may also be helpful in fluid mechanics, but any relevant diagram will be helpful.

(d) Do You Need *Mass Continuity*?

This is
$$\dot{V} = \dot{m}v = uA,$$
where the symbols have the meaning listed in the nomenclature (pp. viii–x).

(e) Do You Need the *Energy* Equation?

This question is bound up with (a) and (b) above. For example, Bernoulli's equation may well be an essential part of the solution since this is a statement of the conservation of mechanical energy. However, for compressible fluid flow, where temperature proves to be a significant variable, the full steady-flow energy equation will be required since this shows the conservation of the sum of both mechanical and thermal energies.

(f) Do You Need the *Momentum* Equation?

As in *Work Out Engineering Thermodynamics*, I have included a reasonably full treatment of this, since students experience so much trouble with it on first meeting it. Momentum ideas are involved whenever forces are to be calculated, whether from the application of fluid pressure, shear forces and body forces or from acceleration or deceleration of fluids or both.

(g) Have You Used the Correct *Language*?

For example, w for specific work transfer (in kJ/kg), W for work transfer (in kJ) and $\dot{W} = dW/dt$ = rate of doing work (or power) (kW), since the dimensions of a differential coefficient are determined by drawing a line, viz. $\dfrac{d}{d}\bigg|\dfrac{W}{t}$, and considering the dimensions to the right of the vertical line i.e. in this case W/t in kJ/s or kW.

(h) Have You Ensured Correct *Dimensional Reasoning*?

This is, of course, allied to (g) above since correct language goes hand in hand with correct dimensions.

Example 1

From hydrostatics we have the hydrostatic equation:

$$p = \rho g h$$

and for $\rho = 10^3 \ \dfrac{kg}{m^3}$; $g = 9.81 \ \dfrac{m}{s^2}$; $h = 2$ m;

i.e. $p = 10^3 \ \dfrac{kg}{m^3} \times 9.81 \ \dfrac{m}{s^2} \times \ 2$ m $= 19\,620 \ \dfrac{kg}{m \ s^2}$,

which is not a readily recognisable unit of pressure.

However, by using unity brackets we can ensure that the answer comes out in obvious units of pressure. Thus in Newton's law we have

$$1 \text{ N} = 1 \text{ kg} \times 1 \text{ m/s}^2$$

or $\left[\dfrac{\text{N s}^2}{\text{kg m}} \right] = \left[\text{UNITY} \right] = \left[\dfrac{\text{kg m}}{\text{N s}^2} \right]$,

and anything multiplied by unity (or its reciprocal) is unchanged.

Thus in the above calculation we may write

$$p = 19\,620 \ \dfrac{kg}{m \ s^2} \left[\dfrac{\text{N s}^2}{\text{kg m}} \right] = 19\,620 \ \dfrac{\text{N}}{\text{m}^2} = 19\,620 \text{ pascals}$$
$$= \qquad 0.1962 \text{ bar.}$$

This gives the correct answer in the right units even though there is no numerical change in this instance. In the next example the necessary introduction of two unity brackets does bring about a substantial change in the numerical answer in order that the right units may appear.

Example 2

From the steady flow of water through a horizontal nozzle from negligible velocity and a pressure of 50 bar to an exit pressure of 40 bar where $\rho = 10^3$ kg/m^3, we have from Bernoulli's equation for incompressible flow:

$$\frac{p_1}{\rho} + \frac{u_1^2}{2} = \frac{p_2}{\rho} + \frac{u_2^2}{2},$$

and since u_1 is given as very small then

$$u_2 = \sqrt{\frac{2(50 - 40) \text{ bar}}{10^3 \ \dfrac{kg}{m^3}}} = \sqrt{\frac{2}{10^2} \ \frac{\text{bar m}^3}{\text{kg}}},$$

which is not recognisable as a velocity.

Now, by introducing two unity brackets we rationalise the result. Thus

$$u_2 = \sqrt{\frac{2}{10^2} \ \frac{\text{bar m}^3}{\text{kg}} \left[\frac{10^5 \text{ N}}{\text{bar m}^2} \right] \left[\frac{\text{kg m}}{\text{N s}^2} \right]} = \sqrt{2000 \ \frac{\text{m}^2}{\text{s}^2}} = 44.72 \ \frac{\text{m}}{\text{s}}.$$

(Note that without dimensions the answer would have come out as 0.1414 with a guess at the dimensions!)

Note: Although the SI unit of pressure is the pascal (Pa) ($= 1$ N/m^2), in this volume use is made of the bar ($= 10^5$ N/m^2) and its multiples and submultiples because this is such common practice elsewhere, for example the millibar in weather forecasting.

Throughout this volume the fundamental ideas are given emphasis in the text to make them stand out and in the two examination papers at the end an attempt is made to apply the above questions wherever possible to show their recurring relevance.

Some General Definitions in Fluid Mechanics

Ideal Fluid

This is assumed to have zero viscosity (for a fuller definition see Chapter 1). There is no conversion of mechanical energy into thermal energy with an incompressible ideal fluid. If an ideal fluid is initially at rest, it can be shown that, subsequently, all particles must continue to have the same total mechanical energy.

Real Fluid

Viscosity is unavoidable in a real fluid; flow of a real fluid involves the conversion of mechanical energy to thermal energy.

Laminar Flow (or Streamline or Viscous Flow)

In this case, fluid particles move along in layers or laminae with one layer sliding over an adjacent layer. The flow is governed by Newton's law of viscosity (for one-dimensional flow). (See Chapter 1 for further treatment.) The degradation in mechanical energy varies approximately as (velocity)^1.

Turbulent Flow

This occurs when the fluid particles move in very irregular paths causing an exchange of momentum from one portion of a fluid to another. The degradation in mechanical energy varies approximately as (velocity)^2.

Laminar flow tends to occur when the fluid velocity is small or the fluid viscosity is large or both. In cases where the given flow can be either laminar or turbulent, the turbulent flow sets up greater shear stresses and causes more mechanical energy to be converted to thermal energy.

Steady Flow

This is characterised by a steady mass-flow rate and by the fact that across any section at right angles to the flow all properties are constant with respect to time. True steady flow is found only in laminar flow.

Steady, turbulent flow is said to exist when the *mean* velocity of flow at a section remains constant with time.

Unsteady Flow

This occurs when conditions at any point change with time. An example is the flow of a liquid being pumped through a fixed system at an increasing rate.

Uniform Flow

This occurs when, at every point, the velocity vector is identical in magnitude and direction at any given instant. That is, $\partial u/\partial s = 0$, in which time is held constant and s is the displacement in any direction. This equation states that there is no change in the velocity vector in any direction throughout the fluid at any instant. It states nothing about the change in velocity at a point with time.

In the flow of a real liquid in an open or closed conduit, the definition may be extended in most cases, even though the velocity vector at the boundary is always zero. When all parallel cross-sections through the conduit are identical (i.e. when the conduit is prismatic), and the average velocity at each section is the same at any given instant, the flow is said to be uniform.

An example is the flow of a liquid being pumped through a long straight pipe (uniform in the direction of flow).

Non-uniform Flow

This occurs when the velocity vector varies from place to place at any instant, i.e. $\partial u/\partial s \neq 0$. An example is the flow of a liquid through a tapered or curved pipe.

Other Examples

Steady, uniform flow   The flow of a liquid through a straight horizontal pipe at a constant rate.

Unsteady, uniform flow   The flow of a liquid through a straight horizontal pipe at a changing rate.

Steady, non-uniform flow   The flow of a liquid through a conical pipe at a constant rate.

Unsteady, non-uniform flow   The flow of a liquid through a conical pipe at a changing rate.

Path Line

This is the path of a single fluid particle taken over a length of time. Path lines may cross each other.

Streamline

This is a curve which is everywhere parallel to the direction of flow. No streamlines can ever cross each other. Streamlines and path lines become identical if the flow does not change with time, i.e. if the flow is steady. Streamline pictures allow fluid flow to be visualised.

Stream Tube

The total flow can be divided into parts, each taking place along an imaginary tube, called a stream tube. A stream tube is bounded by streamlines and hence no flow can cross the boundary of the stream tube. If the fluid is ideal (i.e. if there is no energy degraded) then the total mechanical energy per unit mass of fluid remains constant along a stream tube (i.e. the flow is isenergic).

The product of velocity and cross-sectional area is assumed constant for all stream tubes when drawing streamline fields for an incompressible fluid.

Boundary Layer

When a real fluid flows past a solid boundary, the fluid in the immediate neighbourhood of the boundary is retarded by viscous shear forces set up within the fluid. The retarded layer of fluid is called the boundary layer.

Continuum

There are two ways of approaching problems in fluid mechanics.

In the *microscopic* approach the individual molecular motion is examined mathematically.

In the *macroscopic* approach (used in this volume) the behaviour of the fluid is described mathematically by examining the motion of small volumes of the fluid which contain a large number of molecules. This implies that since we are taking average effects of many molecules the fluid is being regarded as a continuum (i.e. as a continuous medium). This is justified if the characteristic dimension is very much greater than the mean free path of the molecules.

1 Fundamental Properties of a Real Fluid

1.1 Introduction

Fluids are substances which flow, that is, they are matter in a shape so readily distortable that it cannot be said to have a shape of its own. Unlike solids, in which a given shear stress produces a definite change of shape, in fluids the smallest shear stress if applied long enough will produce any required change of shape.

Thus the fundamental mechanical property of a fluid is that in a fluid at rest there can be no shear stress.

Fluids may be divided into two categories: (a) liquids, (b) gases.

Liquids are practically incompressible; they possess a definite size but take the shape of the containing vessel.

Gases are readily compressible; they adopt both the size and shape of the containing vessel, and, if free to expand, do so indefinitely.

Density

Symbolised by ρ, this is mass per unit volume having dimensions M/L^3 (units kg/m^3).

Specific Gravity (S)

This is the ratio of a fluid's density to that of some standard substance or fluid. For liquids this standard is water, usually at $4\,°C$.

Compressibility

The degree to which a fluid offers resistance to compression is expressed by what is known as its bulk modulus of elasticity, K. Let a quantity of fluid at pressure p and volume V be subjected to an increase of pressure dp. Let this increase cause a change of volume $-dV$; then

$$\text{bulk modulus of elasticity, } K = \frac{\text{increase of pressure}}{\text{volumetric strain}}$$

$$= \frac{dp}{-\dfrac{dV}{V}} = -V\frac{dp}{dV}.$$

Because volumetric strain is dimensionless, K will have units of pressure, i.e. F/L^2 or M/Lt^2 (since $F = ML/t^2$). Thus for water $K = 2 \times 10^6 \text{ kN/m}^2$.

The reciprocal of K is a direct measure of the compressibility of the fluid.

Surface Tension

The apparent tension effects which occur on the surfaces of liquids, when the surfaces are in contact with another fluid or solid, depend fundamentally upon the relative sizes of the intermolecular cohesive and adhesive forces. On a liquid surface in contact with the atmosphere, surface tension manifests itself as an apparent 'skin' over the surface, which will support small loads. The surface tension, σ, is the force in the liquid surface normal to a line of unit length drawn in the surface:

$$\sigma = \frac{\text{force}}{\text{unit length}} \qquad \text{and has dimensions } F/L \text{ or } M/t^2,$$

e.g. water at 20 °C has a surface tension of 0.073 N/m.

Vapour Pressure

All liquids tend to vaporise, i.e. change from the liquid to the gaseous state. If the space above the liquid is confined then the molecules of gas escaping from its surface exert their own partial pressure, known as the vapour pressure of the liquid. Vapour pressure increases with temperature.

Boiling occurs when the pressure above the liquid equals or is less than the vapour pressure.

For example, at 20 °C water has a vapour pressure of 2337 N/m^2
and at 100 °C has a vapour pressure of 101 325 N/m^2.

Viscosity

A fluid cannot sustain shear forces in a static condition. However, when a shear force exists and flow takes place then the rate at which the fluid yields to the force varies for different fluids. The property of a fluid by virtue of which it can resist shear forces when in motion is called viscosity.

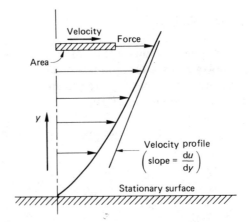

Newton's law of viscosity states that the shear stress is directly proportional to the rate of shear strain (velocity gradient) with the constant of proportionality defined as the coefficient of absolute viscosity; i.e.

$$\tau = \mu \frac{\mathrm{d}u}{\mathrm{d}y} \qquad \text{(for straight, parallel motion)},$$

where $\tau = \dfrac{\text{force}}{\text{area}} = $ shear stress,

$\dfrac{\mathrm{d}u}{\mathrm{d}y} = $ velocity gradient or rate of shearing strain, and

$\mu = $ coefficient of absolute viscosity.

(*Note:* In solids, $\tau = G\, \mathrm{d}x/\mathrm{d}y$, i.e. shear stress $\propto$ shear strain.)

Thus $\mu = \tau \Big/ \dfrac{\mathrm{d}u}{\mathrm{d}y}$ which expressed dimensionally is $\dfrac{F}{L^2} \Big/ \dfrac{L}{tL} = \dfrac{Ft}{L^2}$.

'No slip' Condition

The effect of viscosity is such that fluid particles immediately adjacent to a solid surface do not move with respect to that surface.

Typical units are N s/m^2, centipoise (cP) (= 10^{-3} N s/m^2).

Alternatively, since $F = ML/t^2$, the dimensions of viscosity can be expressed as $ML/t^2 \times t/L^2 = M/Lt$ with typical units kg/(m s).

Newtonian Fluid

This is a fluid which obeys Newton's law of viscous flow. That is, μ is constant with respect to the rate of shear $\mathrm{d}u/\mathrm{d}y$ for a given temperature and pressure.

Kinematic Viscosity

Symbolised by ν, this is defined as the ratio of the absolute viscosity to the fluid density;

i.e.
$$\nu = \frac{\mu}{\rho} = \frac{M}{Lt} \times \frac{L^3}{M} = \frac{L^2}{t}.$$

Typical units are m^2/s or stokes (cm^2/s).

Empirical units are in common commercial use and refer to the time the fluid takes to flow through a standard orifice; thus in Britain Redwood seconds are commonly used, and in North America Saybolt seconds.

These units may be converted to m^2/s or stokes using empirical relationships.

Ideal Fluid

This is a fluid which has zero viscosity, surface tension and vapour pressure effects. An ideal liquid is also incompressible.

The concept of an ideal fluid is useful in the solution of many fluids problems.

Rheological Diagram

This is a graphical plot of τ, the viscous stress, against the differential coefficient du/dy. The diagram shows both Newtonian fluids (a straight line through the origin) and three types of non-Newtonian fluids for comparison.

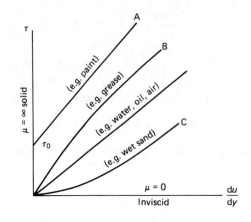

A: $\qquad \tau = \tau_0 + \mu_0 \dfrac{du}{dy}$. $\qquad$ Fluid behaves as a solid till yield stress τ_0 is reached and then as a Newtonian fluid.

B: $\qquad \tau = K \left(\dfrac{du}{dy}\right)^n \qquad (n < 1)$.

C: $\qquad \tau = K \left(\dfrac{du}{dy}\right)^n \qquad (n > 1)$.

1.2 Worked Examples

Example 1.1

The diagram shows two plates Δy apart, the lower one fixed and the upper one free to move under the action of a mass of 25 g as shown. If the fluid between the plates is castor oil (absolute viscosity 650×10^{-3} N s/m^2) and the area of contact of the upper plate with the oil is 0.75 m^2, find the velocity of the upper plate when the distance separating the plates is 1 cm.

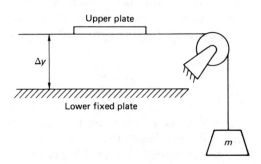

Answer

Newton's law of viscosity: $\tau = \mu \dfrac{du}{dy}$ and with a constant τ,

$$\frac{du}{dy} = \text{constant} = \frac{\Delta u}{\Delta y}.$$

Hence

$$\tau = \mu \frac{\Delta u}{\Delta y}.$$

Gravitational force on mass $m = mg = 25 \text{ g} \left[\dfrac{\text{kg}}{10^3 \text{ g}}\right] 9.81 \dfrac{\text{m}}{\text{s}^2}$.

Viscous shear stress $\tau = \dfrac{\text{shear force}}{\text{area}} = \dfrac{\dfrac{25^3}{10} \text{ kg} \times 9.81 \dfrac{\text{m}}{\text{s}^2}}{0.75 \text{ m}^2} \left[\dfrac{\text{N s}^2}{\text{kg m}}\right]$

$$= 0.327 \frac{\text{N}}{\text{m}^2}.$$

Thus

$$\Delta u = \frac{\tau}{\mu} \Delta y = \frac{0.327 \dfrac{\text{N}}{\text{m}^2} \times 10^3 \times \text{m}^2 \times 1 \text{ cm} \left[\dfrac{\text{m}}{10^3 \text{ cm}}\right]\left[\dfrac{10^3 \text{ mm}}{\text{m}}\right]}{650 \text{ N s}},$$

i.e. for fluid $\qquad\qquad \Delta u = 0.503 \dfrac{\text{mm}}{\text{s}}.$

(*Note:* At lower plate $u = 0$ because of the 'no slip' condition.)

Also plate velocity = $0.503 \dfrac{\text{mm}}{\text{s}}$ because of the 'no slip' condition.

Example 1.2

The diagram shows a plate of negligible weight moving upwards under the action of a force F when situated equidistantly between two fixed surfaces as shown. Kerosene fills the spaces on both sides of the moving plate and has an absolute viscosity of 1.64×10^{-3} N s/m^2. If the contact area on both sides of the moving plate is 2.5 m^2, find the value of the force on the plate if its velocity is constant at 2.5 mm/s.

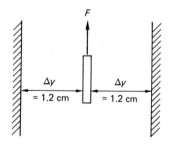

Answer

Total area of contact for plate = $2 \times 2.5 \text{ m}^2 = 5 \text{ m}^2 \qquad (= A)$;

$$\tau = \frac{F}{A} = \mu \frac{du}{dy} = \mu \frac{\Delta u}{\Delta y},$$

$$\text{or} \quad F = A\mu \frac{\Delta u}{\Delta y} = \frac{5 \text{ m}^2 \times 1.64 \times 10^{-3} \dfrac{\text{N s}}{\text{m}} \times 2.5 \dfrac{\text{mm}}{\text{s}} \left[\dfrac{\text{m}}{10^3 \text{ mm}}\right]}{1.2 \text{ cm} \left[\dfrac{\text{m}}{10^2 \text{ cm}}\right]}$$

$$= 17.083 \times 10^{-4} \text{ N}.$$

Example 1.3

In the diagram the total space between the stationary boundaries is 12.7 mm. A plate of infinite dimensions is pulled upwards between ethylene glycol fluid on the left-hand side and propylene glycol fluid on the right-hand side. Find the lateral position of the plate when it finds its equilibrium position, if the thickness of the plate is 0.7937 mm. $\mu_1 = \dfrac{16.2}{10^3} \dfrac{\text{N s}}{\text{m}^2}$; $\mu_2 = \dfrac{42}{10^3} \dfrac{\text{N s}}{\text{m}^2}$.

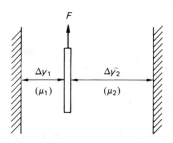

Answer

When the plate finds its equilibrium lateral position the shear stress on each side is the same.

i.e. $\quad \mu_1 \dfrac{\Delta u}{\Delta y_1} = \mu_2 \dfrac{\Delta u}{\Delta y_2} \qquad (\Delta u$ is same both sides$)$.

Thus $\quad \dfrac{\Delta y_2}{\Delta y_1} = \dfrac{\mu_2}{\mu_1} = \dfrac{42}{16.2} = 2.593$.

But $\quad \Delta y_1 + \Delta y_2 = 12.7 \text{ m} - 0.7937 \text{ mm} = 11.9063 \text{ mm}$,

$\qquad \Delta y_1 + 2.593 \, \Delta y_1 = 11.9063 \text{ mm}$,

$\qquad \Delta y_1 = \dfrac{11.9063}{2.593} = 4.592 \text{ mm}$,

and $\quad \Delta y_2 = 11.9063 - 4.592 = 7.314 \text{ mm}$.

Example 1.4

The pressure exerted on a liquid increases from 500 to 1000 kN/m². The volume decreases by 1 per cent. Determine the bulk modulus of the liquid.

Answer

$$\text{Bulk modulus } K = -V \frac{\mathrm{d}p}{\mathrm{d}V} = \frac{\mathrm{d}p}{-\dfrac{\mathrm{d}V}{V}} = \frac{\Delta p}{-\dfrac{\Delta V}{V}} .$$

Now $\Delta p = (1000 - 500) = 500 \dfrac{\text{kN}}{\text{m}^2}$,

and $\dfrac{\Delta V}{V} = -0.01$ (decrease in volume).

Thus $K = \dfrac{500 \dfrac{\text{kN}}{\text{m}^2}}{-(-0.01)} = 5 \times 10^4 \dfrac{\text{kN}}{\text{m}^2}$ (= 0.5 bar).

Example 1.5

Water at $20\,^{\circ}\text{C}$ has a bulk modulus of 21.8×10^8 N/m². Find the pressure rise required to decrease its volume by 1 per cent.

Answer

$$\Delta p = K \left(-\frac{\text{d}V}{V} \right) = 21.8 \times 10^8 \ \frac{\text{N}}{\text{m}^2} \times -(-0.01)$$

$$= 21.8 \times 10^6 \ \frac{\text{N}}{\text{m}^2} = 218 \text{ bar}.$$

2 Hydrostatic Pressure

2.1 Introduction

In this chapter there are two important fundamentals to be examined.

(a) Pascal's principle which states that the pressure at a point in a fluid at rest is the same in all directions.
(b) The hydrostatic equation which gives the value for the pressure at a given depth in a continuous mass of fluid relative to some reference pressure.

These two principles lead to two applications of importance to mechanical engineers:

(a) the science of manometry,
(b) the calculation of forces on immersed surfaces.

Pascal's Principle

Consider the small element of fluid of unit width normal to the page taken at a point in a fluid at rest (see diagram).

In the absence of shear forces the only forces are normal and gravitational forces.

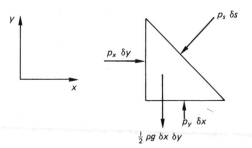

For equilibrium in the x-direction: $p_x\, \delta y - p_s\, \delta s \sin \theta = 0$.

For equilibrium in the y-direction: $p_y\, \delta x - p_s\, \delta s \cos \theta - \rho g\, \dfrac{\delta x\, \delta y}{2} = 0$.

Now $\delta y = \delta s \sin \theta$, $\delta x = \delta s \cos \theta$, and taking the limit as the element reduces to a point (remembering the continuum approach described in the section on general definitions), we get

$$p_x = p_y = p_s \qquad \text{(Pascal's equation).}$$

Furthermore, we can show that two points in the same horizontal plane in a continuous mass of fluid at rest have the same pressure.

The Hydrostatic Equation

This gives the relative pressure at a given depth in a continuous mass of fluid at rest as a simple function of fluid density and depth as shown below.

Refer to the diagram below and consider the small element of fluid at depth h below the free surface on which the reference pressure is acting; then:

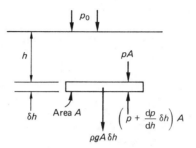

For equilibrium,

$$pA - \left(p + \frac{dp}{dh}\, \delta h\right) A + \rho g A\, \delta h = 0$$

or
$$\frac{dp}{dh} = \rho g;$$

and, for an incompressible fluid, ρ is constant; and integrating with respect to h we get

$$p = \rho g h \qquad \text{(hydrostatic equation).}$$

Note that this is a *gauge* pressure or pressure relative to the free-surface reference pressure p_0.

The *absolute* pressure at depth h is given by

$$p_{\text{abs}} = p_0 + p_{\text{rel}} = p_0 + \rho g h.$$

Note that very often (but not always), $p_0 = p_{\text{atm}}$.

2.2 Manometry

Barometer

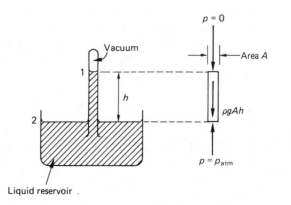

Considering the free-body diagram for the liquid column and summing forces in the vertical direction,

$$\Sigma F = 0 = p_{atm} A - \rho g A h,$$

where upward forces are positive.
Thus
$$p_{atm} = \rho g h$$

(which is the hydrostatic equation again), and for the case where the liquid is mercury and $h = 760$ mm,

$$p_{atm} = 13.6 \times 10^3 \ \frac{kg}{m^3} \times 9.81 \ \frac{m}{s^2} \times 760 \ mm \left[\frac{m}{10^3 \ mm} \right] = 101.396 \ kPa$$
$$(= 1.013 \ 96 \ bar).$$

Manometric Pressure

The barometer analysis shows that vertical columns of liquid can be used to measure pressure. The science of this measurement is called manometry.

There are different types of manometers with varying degrees of sensitivity which embody the principles already derived and are used for pressure measurement. Three examples are given in section 2.4 and others are considered in the exercises at the end of this chapter.

Simple U-tube Manometer

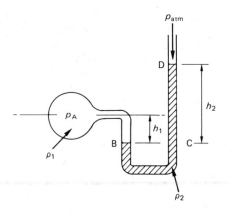

Consider the U-tube manometer connected via a small hole to a pipe carrying a fluid of density ρ_1 at pressure p_A (which is to be measured). Let the open end of the U-tube be subjected to atmospheric pressure, p_{atm}.

At the common surface B–C with the configuration as shown in the diagram we have:

$$p_A + \rho_1 g h_1 = p_B = p_C = p_D + \rho_2 g h_2,$$

or
$$p_A + \rho_1 g h_1 = p_D + \rho_2 g h_2.$$

Now p_A is the pressure to be measured (p) and $p_D = p_{atm}$.
Thus

$$p - p_{atm} = (\rho_2 h_2 - \rho_1 h_1) g.$$

Differential Manometer

This is used to measure the pressure differential between two fluid reservoirs.

$$p_A + \rho_1 g h_1 = p_D + \rho_3 g h_3 + \rho_2 g h_2$$

or differential pressure is given by

$$p_A - p_D = (\rho_3 h_3 + \rho_2 h_2 - \rho_1 h_1)g.$$

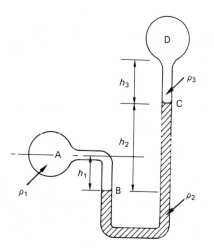

Inverted U-tube Manometer

Another type of differential manometer.

$$p_B + \rho_1 g h_1 = p_A,$$
$$p_B + \rho_2 g h_2 = p_C,$$
$$p_C + \rho_3 g h_3 = p_D;$$

or

$$p_D - \rho_3 g h_3 - \rho_2 g h_2 + \rho_1 g h_1 = p_A,$$

or

$$p_A - p_D = (\rho_1 h_1 - \rho_2 h_2 - \rho_3 h_3)g.$$

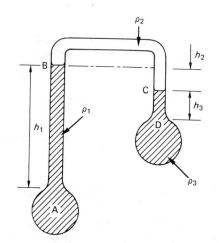

Pressure Gauge

A Bourdon tube pressure gauge has at its base a housing suitable for fitting to a source of pressure to be measured. The pressure to be sensed is connected to a curved tube of elliptical cross-section inside the housing and connected at the other end of the tube to a rack-and-pinion assembly. The applied pressure causes the tube to straighten and the shaft of the assembly carries a needle which is rotated against a calibrated scale giving a measure of the pressure relative to the prevailing atmosphere. The scale is graduated at zero when the tube is exposed to purely atmospheric pressure, and thus the device is called a pressure gauge. It reads the difference in pressure across the Bourdon tube between the fluid and the atmosphere.

2.3 Hydrostatic Forces on Submerged Surfaces

Plane Surfaces

A plane surface of area A is inclined to the horizontal at an angle θ in a fluid of uniform density ρ. O is the point of intersection of the free surface of the fluid with the projection of the inclined plane (see diagram).

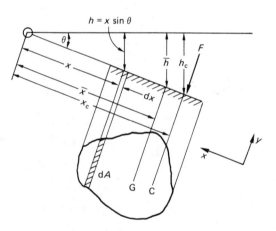

Consider an elemental strip of area dA at a depth h (= $x \sin \theta$) from the fluid surface.

The pressure on dA will be constant and equal to ρgh.

The force on dA normal to the surface is

$$dF = p \, dA = \rho gh \, dA = \rho gx \sin \theta \, dA.$$

The total force over area A is

$$F = \int_0^A \rho gx \sin \theta \, dA = \rho g \sin \theta \int_0^A x \, dA.$$

But $\int_0^A x \, dA$ is termed the first moment of area A about O.

If G denotes the position of the centroid of the area and OG = $\bar{x}$ then by definition the first moment of area $A = A\bar{x}$.

Thus

$$F = \rho g \sin \theta \int_0^A x \, \mathrm{d}A = \rho g A \overline{x} \sin \theta = \rho g A \overline{h}.$$

Note that this is the force on the upper face of the plane only – the force on the lower face depends on the pressure there.

Furthermore, for an irregular shape such as that shown, it is also necessary to take moments in the y-direction to locate the point C. $\overline{h}$ is the vertical depth of the centroid.

Example

To find the hydrostatic thrust on a vertical rectangular plane surface with its upper edge in the free surface of the fluid (see diagram):

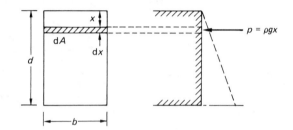

$\sin \theta = 1;$ $x = h.$

Either

$$F = \rho g \int_0^A x \, \mathrm{d}A$$

$$= \rho g \int_0^A xb \, \mathrm{d}x$$

$$= \rho g b \left[\frac{x^2}{2} \right]_0^d = \frac{\rho g b d^2}{2}$$

or

$$F = \rho g A \overline{h} = \rho g A \overline{x}$$

$$= \rho g \, (bd) \, \frac{d}{2} = \frac{\rho g b d^2}{2} .$$

Centre of Force (or Centre of Pressure)

This is the point on the submerged surface at which the resultant hydrostatic force may be considered to act. Its position may be found by summing the moments of forces on elemental strips about O and equating to the moment of the resultant force.

C denotes the position of the centre of force in the upper diagram of page 12, distance x_C from O at a depth of h_C.

The moment of force acting on area $\mathrm{d}A$ about O is given by

$$\mathrm{d}M = x \, \mathrm{d}F = \rho g x^2 \sin \theta \, \mathrm{d}A.$$

13

Thus
$$M = F \times x_C = \int_0^A \rho g x^2 \sin \theta \; \mathrm{d}A = \rho g \sin \theta \int_0^A x^2 \; \mathrm{d}A,$$

$$x_C = \frac{\rho g \sin \theta \int_0^A x^2 \; \mathrm{d}A}{F} = \frac{g \sin \theta \int_0^A x^2 \; \mathrm{d}A}{g \sin \theta \int_0^A x \; \mathrm{d}A}$$

$$= \frac{\int_0^A x^2 \; \mathrm{d}A}{\int_0^A x \; \mathrm{d}A}$$

But $\int_0^A x^2 \; \mathrm{d}A$ is termed the second moment of area about O and is written I_O.

Hence $x_C = \dfrac{\text{second moment of area about O}}{\text{first moment of area about O}} = \dfrac{I_O}{A\bar{x}}.$

By the parallel axes theorem, $\qquad I_O = I_G + A\bar{x}^2,$

giving $\qquad x_C = \dfrac{I_G + A\bar{x}^2}{A\bar{x}} = \bar{x} + \dfrac{I_G}{A\bar{x}} = \bar{x} + \dfrac{k_G^2}{\bar{x}}.$

$\left(\textit{Note:} \; k_G = \text{radius of gyration and } h_C = x_C \sin \theta = \bar{x} \sin \theta + \dfrac{I_G \sin \theta}{A\bar{x}} \right.$

$\qquad\qquad \text{or} \quad h_C = \bar{h} + \dfrac{I_G \sin^2 \theta}{A\bar{h}} \Bigg).$

Example

For the vertical rectangular plane surface in the diagram on page 13 we get: $\sin \theta = 1$;

$$x_C = h_C = \frac{\int_0^A x^2 \; \mathrm{d}A}{\int_0^A x \; \mathrm{d}A} = \frac{\int_0^d x^2 b \; \mathrm{d}x}{\int_0^d xb \; \mathrm{d}x} = \frac{\left[b\frac{x^3}{3} \right]_0^d}{\left[b\frac{x^2}{2} \right]_0^d} = \frac{b\frac{d^3}{3}}{b\frac{d^2}{2}} = \frac{2}{3}d$$

or $\qquad h_C = \bar{h} + \dfrac{I_G \sin^2 \theta}{A\bar{h}} = \bar{x} + \dfrac{I_G}{A\bar{x}} = \dfrac{d}{2} + \dfrac{\frac{bd}{12}}{(bd)\frac{d}{2}} = \dfrac{d}{2} + \dfrac{d}{6} = \dfrac{2}{3}d.$

(*Note:* For a rectangle, $I_G = bd^3/12$; for a circle $I_G = \pi d^4/64$.)

Hydrostatic Forces on Curved Surfaces

Consider the diagram. The hydrostatic force acting on an elemental strip of area $\mathrm{d}A$ and depth h is $p \; \mathrm{d}A$ normal to the surface.

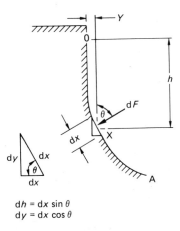

$$dh = dx \sin \theta$$
$$dy = dx \cos \theta$$

But $p = \rho gh$, and if the width of the strip at X is B then

$$dA = B\, dx,$$

$dF = p\, dA = \rho ghB\, dx$ acting at θ to the vertical.

To sum the forces acting over similar elemental strips the horizontal and vertical components must be obtained:

$$dF_{\mathrm{H}} = \rho ghB\, dx \sin \theta = \rho gh\,(B\, dh),$$

$$dF_{\mathrm{V}} = \rho ghB\, dx \cos \theta = \rho gh\,(B\, dy),$$

giving $\qquad F_{\mathrm{H}} = \displaystyle\int_{0}^{A} \rho gBh\, dh \qquad$ and $\qquad F_{\mathrm{V}} = \displaystyle\int_{0}^{A} \rho gBh\, dy.$

But $\displaystyle\int_{0}^{A} Bh\, dh$ = first moment of area of the horizontal projection of surface OA onto a vertical plane

$\qquad\qquad = A\bar{h}$ (where A is the projected area onto a vertical plane), and

$$F_{\mathrm{H}} = \rho gA\bar{h}.$$

Also, $\displaystyle\int_{0}^{A} Bh\, dy$ = volume of fluid above the curved surface OA, and

$$F_{\mathrm{V}} = \rho g \times \text{volume of fluid above OA} = \text{weight of fluid above OA}.$$

The resultant force $F = \sqrt{F_{\mathrm{H}}^{2} + F_{\mathrm{V}}^{2}}$ acting at $\tan^{-1} F_{\mathrm{H}}/F_{\mathrm{V}}$ to the vertical.

Centre of Force

Horizontal component   The centre of force for a hydrostatic force acting on a vertical plane is

$$h_{\mathrm{c}} = \frac{\displaystyle\int_{0}^{A} h^{2}\, dA}{\displaystyle\int_{0}^{A} h\, dA} = \frac{\displaystyle\int_{0}^{A} Bh^{2}\, dh}{\displaystyle\int_{0}^{A} Bh\, dh} = \frac{I_{\mathrm{O}}}{A\bar{h}} = \bar{h} + \frac{I_{\mathrm{G}}}{A\bar{h}}.$$

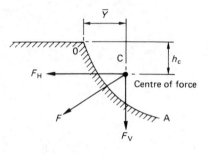

Vertical component

This will act through the centroid of the volume above OA:

$$\bar{y} = \frac{\displaystyle\int_0^A Bhy\ dy}{\displaystyle\int_0^A Bh\ dy}.$$

By combining the components at their points of application the resultant centre of force may be obtained.

2.4 Worked Examples

Example 2.1

In the arrangement shown in the diagram, atmospheric pressure is applied to the top of the right-hand limb and the oil in the pipe and the left-hand limb has a specific gravity of 0.82. Given that the density of water is 10^3 kg/m^3 and the specific gravity of mercury is 13.6, determine the gauge pressure at A.

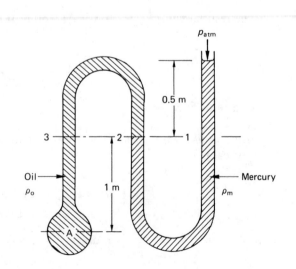

At the common level 1–2–3 all pressures are equal.

$$p_1 = p_2 = p_3 = \rho_m g h_m$$

where h_m is 0.5 m (of mercury above level) and all pressures are relative to p_{atm}.

Thus $p_A = p_1 + \rho_0 g h_2$

where h_2 is 1 m (of oil below level)

or $p_A = 9.81 \frac{m}{s^2} \left[\left(13.6 \times 10^3 \frac{kg}{m^3} \times 0.5 \text{ m}\right) + \left(0.82 \times 10^3 \frac{kg}{m^3} \times 1 \text{ m}\right) \right] \left[\frac{N s^2}{kg \ m} \right]$

$= 74\,828 \ \frac{N}{m^2} = 74.828 \ \frac{kN}{m^2}$ (or 0.74 828 bar)

relative to atmosphere.

Example 2.2

In the arrangement shown in the diagram, the two open-ended limbs are initially subjected to atmospheric pressure with the inside diameter of the tube equal to 7 mm, while the open-ended diameters at A and B are both 44 mm.

Find the pressure difference between A and B due to an increased pressure applied to side B if the surface of separation moves 100 mm. The oil has a specific gravity of 0.83 and the answer should be given in mm of water.

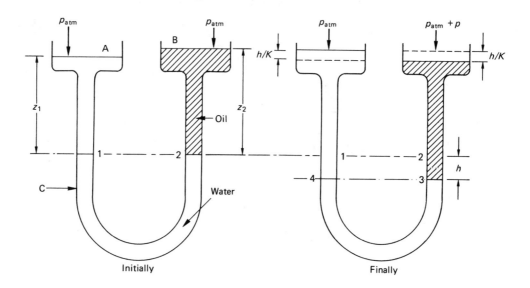

Answer

Let the ratio of the areas of the enlarged ends A and B to the small-bore tube C be K.

Then $K = \left(\dfrac{44}{7}\right)^2 = 39.5$ (since cross-sectional area $\propto d^2$).

Initially at the common level, $p_1 = p_2$,

or $\qquad\qquad\qquad \rho_w g z_1 = \rho_o g z_2,$ (p_{atm} cancels) $\qquad\qquad$ ... (1)

or $\qquad\qquad\qquad z_1 = 0.83 z_2.$

Finally at the new common level, $p_3 = p_4$,

or $\qquad\qquad \rho_w g \left(z_1 + h + \dfrac{h}{K} \right) = \rho_o g \left(z_2 + h - \dfrac{h}{K} \right) + p.$ $\qquad$ ... (2)

Subtracting (1) from (2), we get

$$\rho_w g \left(h + \frac{h}{K} \right) = \rho_o g \left(h - \frac{h}{K} \right) + p$$

or $\qquad\qquad h \left[\left(1 + \dfrac{1}{39.5} \right) - 0.83 \left(1 - \dfrac{1}{39.5} \right) \right] = \dfrac{p}{\rho_w g},$

and for $h = 100$ mm,

$$\frac{p}{\rho_w g} = 100 \text{ mm} (1.0253 - 0.809) = 21.63 \text{ mm water.}$$

Example 2.3

A sensitive manometer has the dimensions shown in diagram (a). The fluid A is oil of specific gravity 0.9 and the fluid B is carbon tetrachloride of specific gravity 1.6.

Find the length of the scale which corresponds to a change of pressure at M of 1 kN/m^2.

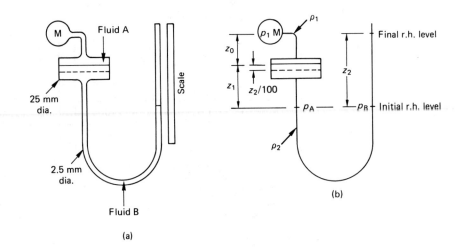

(a)

(b)

$$\text{Area ratio} = \left(\frac{25}{2.5} \right)^2 = 100.$$

$$\text{Pressure increase } p = 1 \, \frac{\text{kN}}{\text{m}^2}.$$

Answer

Initially, $p_A = p_B = 0$,

relative to atmosphere on r.h. limb, or

$$p_1 + \rho_1 g z_0 + \rho_2 g z_1 = 0.$$

Finally, $(p_1 + p) + \rho_1 g \left(z_0 + \dfrac{z_2}{100} \right) + \rho_2 g \left(z_1 - \dfrac{z_2}{100} \right) = \rho_2 z_2 g.$

Subtracting,

$$p + \rho_1 g \, \frac{z_2}{100} + \rho_2 g \left(\frac{-z_2}{100} \right) = \rho_2 z_2 g$$

$$p = z_2 g \left(\rho_2 + \frac{\rho_2}{100} - \frac{\rho_1}{100} \right)$$

$$= z_2 g \times 10^3 \ \frac{\text{kg}}{\text{m}^3} \ (1.6 + 0.016 - 0.009),$$

or $z_2 = \dfrac{1 \, \dfrac{\text{kN}}{\text{m}^2}}{9.81 \, \dfrac{\text{m}}{\text{s}^2} \times 10^3 \, \dfrac{\text{kg}}{\text{m}^3} \times 1.607} \left[\dfrac{\text{kg m}}{\text{N s}^2} \right] \left[\dfrac{10^3 \text{ mm}}{\text{m}} \right] \left[\dfrac{10^3 \text{ N}}{\text{kN}} \right]$

$$= 63.4 \text{ mm}.$$

Example 2.4

For the manometer shown in the diagram, (a) determine an expression for the gauge pressure p in the cylinder, in mm water, in terms of y, the rise in fluid in the inclined tube, θ, the inclination of the tube, S, the specific gravity of the fluid, and K, the ratio of the cylinder diameter to the tube diameter. Hence (b), find the value of K such that the error due to disregarding the change in level in the cylinder will not exceed 0.1 per cent when $\theta = 30°$ and (c) state the magnification of the manometer reading achieved by inclining the tube.

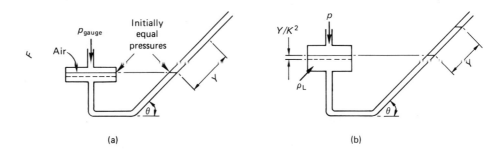

(a)   (b)

Answer

(a) $p = \rho_L g \left(y \sin \theta + \dfrac{y}{K^2} \right) = S \rho_w g y \left(\sin \theta + \dfrac{1}{K^2} \right).$

(b) Neglecting $\dfrac{y}{K^2}$, $p = S \rho_w g y (\sin \theta)$.

Error is $\delta p = S \rho_w g y \left(\dfrac{1}{K^2} \right)$;

percentage error is $\frac{\delta p}{p} = \frac{1}{K^2 \sin \theta} \times 100.$

Thus $0.1 = \frac{100}{K^2 \sin \theta}$

or $K^2 = \frac{1000}{\sin 30°} = 2000$

and $K = 44.7.$

(c) With the tube vertical, $\sin \theta = 1$;

$$p = S\rho_w gy \left(1 + \frac{1}{K^2}\right)$$

$$S\rho_w gy_{\text{vert}} \left(1 + \frac{1}{K^2}\right) = S\rho_w gy \left(\sin \theta + \frac{1}{K^2}\right)$$

or $$\frac{y}{y_{\text{vert}}} = \frac{1 + \frac{1}{K^2}}{\sin + \frac{1}{K^2}} = \frac{1.0005}{0.5005} \approx 2.$$

Example 2.5

A circular plate of diameter 1.5 m is immersed in water, its greatest and least depths below the surface being 1.8 m and 0.9 m respectively.

Find (a) the total force on one face of the plate,

      (b) the position of the centre of force.

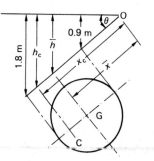

Answer

(a) $$\bar{h} = \frac{1.8 + 0.9}{2} = 1.35 \text{ m.}$$

Force on plate $= \rho g A \bar{h}$

$$= 10^3 \frac{\text{kg}}{\text{m}^3} \times 9.81 \frac{\text{m}}{\text{s}^2} \times \frac{\pi}{4} \times 1.5^2 \text{ m}^2 \times 1.35 \text{ m}$$

$$= 23\,403 \frac{\text{kg m}}{\text{s}^2} \left[\frac{\text{N s}^2}{\text{kg m}}\right] \left[\frac{\text{kN}}{10^3 \text{ N}}\right]$$

$$= 23.4 \text{ kN.}$$

(b)

$$\sin\theta = \frac{1.8 - 0.9}{1.5} = 0.6.$$

$$x_c = \frac{I_O}{A\bar{x}} = \frac{\bar{I} + A\bar{x}^2}{A\bar{x}} = \bar{x} + \frac{k_G^2}{\bar{x}}.$$

$$\bar{x} = \frac{\bar{h}}{\sin\theta} = \frac{1.35\ \text{m}}{0.6} = 2.25\ \text{m}.$$

$$k_G^2 = \frac{I_O}{A} = \frac{(\pi/64)\,d^4}{(\pi/4)\,d^2} = \frac{d^2}{16} = \frac{1.5^2}{16}\ \text{m}^2 = 0.141\ \text{m}^2.$$

$$x_c = 2.25 + \frac{0.141}{2.25} = 2.3125\ \text{m}.$$

$$h_c = 2.3125\sin\theta = 1.388\ \text{m}.$$

Example 2.6

A vertical dock gate is 5.5 m wide and has water to a depth of 7.3 m on one side, and to a depth of 3 m on the other side. Find the resultant horizontal force on the dock gate and the position of its line of action.

To what position does this line tend as the depth of water on the shallow side rises to 7.3 m?

Answer

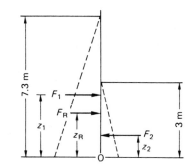

$$F_1 = \rho g\,\Delta\bar{h}$$

$$= 10^3\ \frac{\text{kg}}{\text{m}^3} \times 9.81\ \frac{\text{m}}{\text{s}^2} \times 7.3 \times 5.5\ \text{m} \times \frac{7.3}{2}\ \text{m}\left[\frac{\text{N s}^2}{\text{kg m}}\right]\left[\frac{\text{kN}}{10^3\ \text{N}}\right]$$

$$= 1437.6\ \text{kN},$$

acting at

$$h_{c,1} = \tfrac{2}{3} \times 7.3 = 4.87\ \text{m below free surface.}$$

Similarly,

$$F_2 = 10^3\ \frac{\text{kg}}{\text{m}^3} \times 9.81\ \frac{\text{m}}{\text{s}^2} \times 3 \times 5.5\ \text{m}^2 \times \tfrac{3}{2}\ \text{m}\left[\frac{\text{N s}^2}{\text{kg m}}\right]\left[\frac{\text{kN}}{10^3\ \text{N}}\right]$$

$$= 242.8\ \text{kN},$$

acting at

$$h_{c,2} = \tfrac{2}{3} \times 3 = 2\ \text{m below the free surface.}$$

Thus resultant force F_R = 1437.6 − 242.8 = 1194.8 kN.

Taking moments about O,

$$(F_1 \times z_1) + (F_2 \times z_2) = (F_R \times z_R)$$

or $z_R = \dfrac{[1437.6 \times (7.3 - 4.87)] - [242.8 \times (3 - 2)]}{1194.8} = 2.721$ m above base.

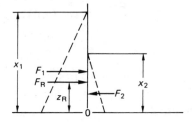

F_1 acts at $\dfrac{x_1}{3}$ above O (i.e. $z_1 = x_1/3$);

F_2 acts at $\dfrac{x_2}{3}$ above O (i.e. $z_2 = x_2/3$).

$$F_1 = \rho g A \overline{h}_1 = \rho g x_1 b \cdot \tfrac{2}{3} x_1,$$

$$F_2 = \rho g A \overline{h}_2 = \rho g x_2 b \cdot \tfrac{2}{3} x_2.$$

Thus $\qquad (F_1 \times z_1) - (F_2 \times z_2) = \tfrac{2}{9} \rho g b\,(x_1^3 - x_2^3) \qquad$ for moments about O.

But, correspondingly, $\qquad (F_1 - F_2)z_R = \tfrac{2}{3} \rho g b\,(x_1^2 - x_2^2)z_R$

or $\qquad z_R = \dfrac{\tfrac{2}{9} \rho g b\,(x_1^3 - x_2^3)}{\tfrac{2}{3} \rho g b\,(x_1^2 - x_2^2)} = \dfrac{(x_1^3 - x_2^3)}{3(x_1^2 - x_2^2)}$

or $\qquad z_R = \dfrac{(x_1^2 + x_1 x_2 + x_2^2)}{3(x_1 + x_2)},$

and, as $x_2 \to x_1$, $\qquad z \to \dfrac{3x_1^2}{3 \times 2x_1} \to \dfrac{x_1}{2}.$

Example 2.7

A hollow triangular box, the ends of which are equilateral triangles, is submerged in water so that one of its rectangular faces lies in the surface of the water. Derive expressions for the net total force on one of the triangular ends and the position of the centre of force. Evaluate these if the sides of the triangle have a length of 1.2 m.

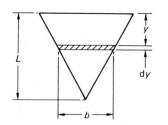

Answer

Force on elemental strip = pressure × area

$$= \rho g y \times b \, dy.$$

Now

$$\frac{b}{D} = \frac{L - y}{L}.$$

Elemental force $= \dfrac{\rho g y \, (L - y) D \, dy}{L}.$

Total force $= \rho g \dfrac{D}{L} \displaystyle\int_0^L (Ly - y^2) \, dy = \rho g \dfrac{D}{L} \left[\dfrac{Ly^2}{2} - \dfrac{y^3}{3} \right]_0^L.$

$$= \rho g \frac{D}{L} \left(\frac{L^3}{6} \right) = \frac{\rho g D L^2}{6}.$$

Now

$$L = D \cos 30° \qquad \text{and}$$

total force $= \dfrac{\rho g D^3 \cos^2 30°}{6} = \dfrac{10^3 \ (kg/m^3) \times 9.81 \ (m/s^2) \times 1.2^3 \ m^3 \times 0.866^2}{6} \left[\dfrac{N \ s^2}{kg \ m} \right]$

$$= 2119 \ N.$$

Depth of centre of force $= \dfrac{\text{moment of total force about surface}}{\text{total force}}$

$$= \frac{\rho g \dfrac{D}{L} \displaystyle\int_0^L (Ly - y^2) y \, dy}{\frac{1}{6} \rho g D L^2} = \frac{6}{L^3} \left[\frac{Ly^3}{3} - \frac{y^4}{4} \right]_0^L$$

$$= \frac{L^4}{12} \times \frac{6}{L^3} = \frac{L}{2} = \frac{D \cos 30°}{2} = \frac{1.2 \ m \times 0.866}{2}$$

$$= 0.52 \ m.$$

Example 2.8

The gate shown in the diagram is pinned at O. Find the height of the free surface, h, at which the gate will start to rotate, neglecting the weight of the gate.

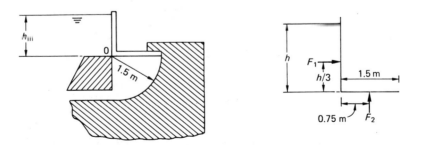

Answer

$$F_1 = \rho g A_1 \bar{h} = \rho g h \times 1 \times h = \rho g \frac{h^2}{2};$$

$$F_2 = \rho g A_2 h = \rho g h \times 1 \times 1.5 = \rho g h \tfrac{3}{2}.$$

Moments about O:
$$\Sigma M_O = 0 = \rho g \frac{h^2}{2} \times \frac{h}{3} - \tfrac{3}{2} \rho g h \, \tfrac{3}{4}$$

or
$$\frac{h^2}{6} = \tfrac{9}{8}$$

or
$$h = 2.598 \text{ m}.$$

Example 2.9

A hemispherical bowl with a flange is to be cast in a pair of moulding boxes. The bowl is 0.3 m radius outside and 0.285 m radius inside. The flange is 0.75 m diameter overall and is 0.015 m thick. The header, for pouring, continues 0.15 m above the top.

Calculate the upward force due to the molten metal tending to lift the top moulding box when the mould and header are filled with molten cast iron of density 6710 kg/m³. Refer to the diagram.

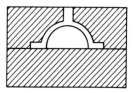

Answer

Referring to the diagram, if the top moulding box is not to move, i.e. if it is to be in equilibrium and at rest, the upward force due to fluid pressure must be equal and opposite to the weight of molten fluid which *could* be contained in the shaded volume which acts downwards due to gravity.

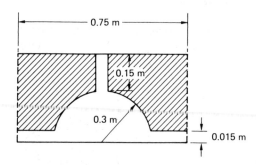

Neglecting the volume of the rises as being small compared with the rest,

$$F_V = \rho g \left[\left(0.45 \times \frac{\pi}{4} \times 0.75^2 \right) - \left(\frac{2}{3} \pi \times 0.3^3 \right) - \frac{\pi}{4} \left(0.75^2 - 0.6^2 \right) 0.015 \right] \text{N};$$

$$F = 6710 \times 9.81 \left[0.199 - 0.0565 - 0.0239 \right] \text{N}$$

$$= 6710 \times 9.81 \times 0.1186 \text{ N}$$

$$= 7807 \text{ N}.$$

Example 2.10

A masonry bridge of parabolic form spans a river. The horizontal distance between the piers on each side of the river is 20 m and the maximum vertical height of the underside of the arch above a horizontal line joining the points on the piers from which the arch springs is 3 m. Taking either of these two points as an origin for rectangular coordinates, obtain the equation of the parabola.

If the water level in the river rises to 1.53 m above the points from which the arch springs, calculate the upthrust on the arch per metre length in the direction of the river flow.

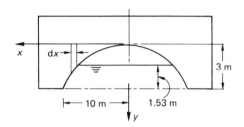

Answer

Equation to a parabolic arch is of the form

$$y = Ax^2 \qquad (A \text{ a constant}).$$

Boundary conditions: when $y = 3$; $x = 10$.

Thus
$$A = \frac{3}{100} = 0.03 \qquad \text{and} \qquad y = \frac{3}{100}x^2 .$$

Now, when $y = (3 - 1.53) = 1.47$, $x = \sqrt{\dfrac{1.47}{0.03}} = 7$.

Thus
$$\text{upthrust} = 2\rho g \int_7^{10} (y - 1.47)\, dx = 2\rho g \int_7^{10} (0.03x^2 - 1.47)\, dx$$

$$= 2\rho g \left[0.010x^3 - 1.47x \right]_7^{10} = 2\rho g \left[(10 - 3.43) - (1.47 \times 3) \right]$$

$$= 2 \times 10^3 \times 9.81\,(6.57 - 4.41)\ \text{N} \qquad \left(\text{with } \rho \text{ in } \frac{\text{kg}}{\text{m}^3} \text{ and } g \text{ in } \frac{\text{m}}{\text{s}^2} \right)$$

$$= 42\,379\ \text{N} = 42.38\ \text{kN}.$$

Example 2.11

The profile of the water face of a dam is given by the equation $44.75y = x^{2.5}$, where y is the height in metres above floor level and x is set back off the face from a vertical line of reference in metres. The depth of water is 40 m.

Specify completely the resultant force due to water pressure per metre of dam, giving its magnitude in MN, its inclination to the vertical and the point where this resultant force cuts a horizontal line at floor level.

25

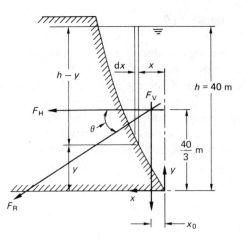

Answer

The horizontal, vertical and resultant forces all emanate from a point $\frac{40}{3}$ m above base. Consider an elemental strip of width dx and of coordinates x (from O) and height $(h - y)$ from water surface to dam profile as shown in the diagram.

Equation of dam wall is:

$$44.75y = x^{2.5}.$$

Boundary conditions: when $y = 40$ m, $x^{2.5} = 44.75 \times 40 = 1790$.
Thus $x = 1790^{0.4} = 20$ m.

Per unit width of wall,

$$F_H = \rho g A \overline{h} = \rho g \times 40 \times 20 \text{ N} \qquad (\rho_1 g \text{ in usual units}),$$

$$F_V = \rho g \int_0^{20} (h - y)\, dx = \rho g \int_0^{20} \left(40 - \frac{x^{2.5}}{44.75}\right) dx$$

$$= \rho g \left[40 - \frac{2x^{3.5}}{7 \times 44.75}\right]_0^{20} = \rho g\,(800 - 228.4)$$

$$= 571.6\rho g \text{ N}.$$

Thus resultant force

$$F_R = \sqrt{F_H^2 + F_V^2} = \rho g \sqrt{800^2 + 571.6^2} = 983.2\rho g \text{ N}$$

$$= 9.645 \text{ MN per metre length}.$$

$$\theta = \arctan \frac{F_V}{F_H} = \arctan \frac{571.6}{800} = 35.55°.$$

To obtain x_0, take moments about O.

$$F_V \times x_0 = \int_0^{20} \rho g\,(h - y)\, x\, dx = \int_0^{20} \rho g \left(40 - \frac{x^{2.5}}{44.75}\right) x\, dx$$

$$= \int_0^{20} \rho g \left(40x - \frac{x^{3.5}}{44.75}\right) dx = \rho g \left[40\frac{x^2}{2} - \frac{2x^{4.5}}{9 \times 44.75}\right]_0^{20}$$

$$= \rho g\,[8000 - 3553.2] = 4446.7\rho g \text{ Nm},$$

or $\qquad x_0 = \dfrac{4446.7\rho g}{571.6\rho g} = 7.779$ m.

Thus $\quad x$ (where F_R cuts the base) $= x_0 + \dfrac{40}{3\tan\theta} = 7.779 + \dfrac{40\times 800}{3\times 571.6}$

$$= 7.779 + 18.66 = 26.44 \text{ m.}$$

Example 2.12

A cylindrical pressure vessel, as shown in the diagram, has a circular opening in the bottom which is blocked by a spring-loaded ball valve, the ball being inside the vessel and the spring beneath the ball. The vessel is filled with oil of specific gravity 0.9 to a depth of 2 m. Compressed air at a pressure of 0.5 bar fills the upper part of the vessel. The ball valve is 0.3 m in diameter, and its specific gravity is 3.6. The circular opening in the bottom of the vessel is such as to support an angle of 120° at the centre of the ball.

Determine the spring force to maintain the vessel just oil-tight.

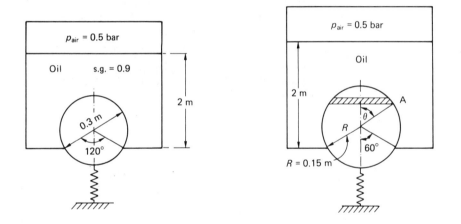

Answer

At A, $\qquad p = 50 \dfrac{\text{kN}}{\text{m}^2} + (2 - 0.15\cos 60° - 0.15\cos\theta)\rho g$

$$= 50 + (2 - 0.075 - 0.15\cos\theta)\rho g$$

$$= (50 + 1.925 \times 0.9g - 0.15 \times 0.9g\cos\theta)\times 10^3 \ \dfrac{\text{N}}{\text{m}^2}$$

$$= (67 - 1.324\cos\theta)\ \dfrac{\text{N}}{\text{m}^2}\times 10^3 .$$

$\text{d}F$ on element $= p\,\text{d}A = p\,2\pi R\sin\theta \cdot R\,\text{d}\theta,$

$\text{d}F_\text{v} = p\,2\pi R\sin\theta \cdot R\,\text{d}\theta \cdot \cos\theta = 2\pi R^2 \sin\theta\cos\theta \cdot p\,\text{d}\theta.$

27

Thus

$$F_V = \int_0^{120} 2\pi R^2 \sin\theta \cos\theta (67 - 1.324 \cos\theta) \, d\theta \times 10^3$$

$$= 2\pi R^2 \left[67 \frac{\sin^2\theta}{2} + 1.324 \frac{\cos^3\theta}{3} \right]_0^{120} \times 10^3$$

$$= 2\pi \times 0.15^2 \left[\frac{67 \times 3}{8} + \frac{1.324}{3} (-\tfrac{1}{8} - 1) \right] \times 10^3$$

$$= 2\pi \times 0.15^2 \, [25.125 - 0.4965] \times 10^3$$

$$= 3481.7 \text{ N.}$$

Weight of sphere $= \tfrac{4}{3} \pi R^3 \rho_s g = \tfrac{4}{3} \pi \times 0.15^3 \times 3600 \times 9.81 \text{ N} = 499.3 \text{ N.}$

$\therefore$ Spring force $= 3481.7 + 499.3 = 3981 \text{ N.}$

Exercises

1 A hydraulic car jack has an input plunger size of 5 mm in diameter and an output jack diameter of 30 mm. It is to be used to raise the rear end of a car of total weight 16 kN. Assuming that the jack will carry half of this, what input force is going to be needed to raise the car rear from the ground?   (*Answer:* 0.222 kN)
2 Determine the pressure difference between A and B in the manometer shown in the diagram below, given that the specific gravities of oil and mercury are respectively 0.88 and 13.6.   (*Answer:* 262.5 kN/m²)

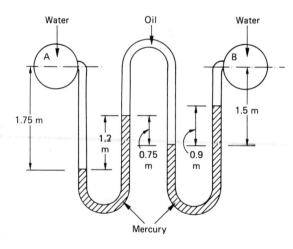

3 The manometer in the diagram (top of p. 29) consists of a small-bore tube with enlarged ends on which atmospheric pressure is bearing. The inside diameter of tube C is 7 mm, and the inside diameters of A and B are both 44 mm. The oil has a specific gravity of 0.83 and a further increment of pressure on side B causes the surface of separation of oil and water to move 100 mm. Find the pressure difference in mm of water between A and B.   (*Answer:* 21.63)

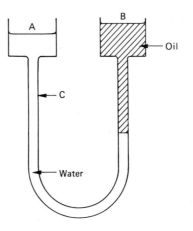

4 A rectangular opening in a vertical face of a dam impounding water is closed by a gate mounted on horizontal trunnions parallel to the longest edge of the gate and passing through the centre of the shorter vertical edges (diagram). If the water level is above the top of the gate, show that the torque required to keep the gate closed is independent of the water level. Determine the magnitude of this torque when the gate is 1.25 m long and 1 m deep. (*Answer:* 1020 N m)

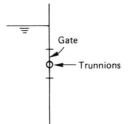

5 Oil floats on water in a tank with vertical sides. The depth of water is 1.8 m and that of the oil is 1.20 m. Calculate the total force per metre run on the vertical wall and the position of the centre of force measured from the base. The oil density is 800 kg/m^3. (*Answers:* 38.5 kN, 0.968 m)

6 The sluice gate shown in the diagram consists of a cylindrical surface AB, 8 m long supported by a hinge at O. Find the magnitude and direction of the total hydrostatic force on the gate. (*Answer:* 0.99 MN, 14.3° inclined to horizontal)

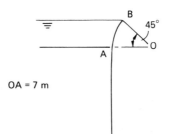

OA = 7 m

7 A gate in the shape of a quadrant of a cylinder is hinged along the top edge and is shown in the diagram. Determine the vertical force necessary to hold the gate in the closed position. The gate is 1 m long and has negligible weight.

(Answer: 49.9 kN)

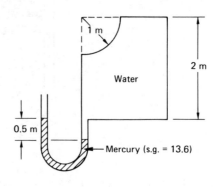

3 The Steady, One-dimensional Flow of an Ideal Incompressible Fluid

3.1 Introduction

The classification of fluid flow can be linked to the manner in which velocity varies in the flow field.

One-dimensional Flow

If the fluid and flow parameters are constant over any cross-section normal to the flow, or if they can be represented by average values over the cross-section, the flow is said to be one-dimensional. An example is given in diagram (a).

(a)

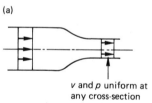

v and p uniform at any cross-section

Two-dimensional Flow

Two-dimensional flow refers to fluid flow where the flow parameters have gradients in two, mutually perpendicular directions. An example is a pipe flow where the velocity at any cross-section is parabolic (diagram (b)).

(b)

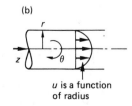

u is a function of radius

Note that we can assume one-dimensional flow in many cases when the flow is strictly two-dimensional to simplify calculations and obtain a more rapid answer. This involves using, for example, an average velocity to replace the parabolic profile shown in diagram (b) and assuming this to be constant over the cross-section. This clearly introduces some error for the sake of speed. In most cases the velocity is zero relative to the solid surface and a parabolic profile is clearly far more representative.

3.2 One-dimensional Equation of Continuity of Flow

Consider the fluid within a streamtube between two sections A and B. Since there can be no flow across the walls of the tube, after a given time δt the same fluid will be contained between A' and B'. Since also under steady conditions of flow the mass of the fluid between A and B remains constant, the mass entering the section at A will equal the mass leaving at B.

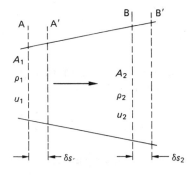

A = Area
ρ = Density
u = Velocity

Thus
$$\rho_1 A_1 \, \delta s_1 = \rho_2 A_2 \, \delta s_2.$$

Dividing by δt,
$$\rho_1 A_1 \, \frac{\delta s_1}{\delta t} = \rho_2 A_2 \, \frac{\delta s_2}{\delta t}.$$

And as $\delta t \to 0$,
$$\frac{\delta s_1}{\delta t} \to u_1, \qquad \frac{\delta s_2}{\delta t} \to u_2.$$

And
$$\rho_1 A_1 u_1 = \rho_2 A_2 u_2 \qquad \text{(the continuity equation)}.$$

For incompressible fluids,
$$\rho_1 = \rho_2.$$

And
$$A_1 u_1 = A_2 u_2 = \dot{V} = \frac{\mathrm{d}V}{\mathrm{d}t}.$$

3.3 Euler's Equation for an Ideal Fluid in One-dimensional Flow

Consider the steady flow of an ideal fluid along a streamtube of elemental area δA. The velocity at any cross-section will be uniform over the section, and, since the fluid is ideal, there will be no viscous shear forces acting over the surface.

32

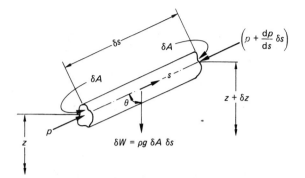

Newton's second law gives:

$$\text{force} = \text{mass} \times \text{acceleration}.$$

Resolving along the streamtube,

$$p\,\delta A - \left(p + \frac{dp}{ds}\,\delta s\right)\delta A \quad \rho g\,\delta A\,\delta s\,\cos\theta = \rho\,\delta A\,\delta s\,a_s,$$

where $a_s = \dfrac{du}{dt}$ (the material or total derivative) $= \dfrac{du}{ds}\,\dfrac{ds}{dt} = u\,\dfrac{du}{ds}$.

Substituting $\cos\theta = \dfrac{dz}{ds}$, we get

$$-\frac{dp}{ds}\,\delta s\,\delta A - \rho g\,\delta A\,\delta s\,\frac{dz}{ds} = \rho\,\delta A\,\delta s\,u\,\frac{du}{ds}.$$

Dividing through by $\rho\,\delta s\,\delta A$ and rearranging,

$$u\,\frac{du}{ds} + \frac{1}{\rho}\,\frac{dp}{ds} + g\,\frac{dz}{ds} = 0 \qquad \text{(Euler's equation)}.$$

Integration of Euler's equation along a streamline:

$$\frac{u^2}{2} + \int\frac{dp}{\rho} + gz = \text{a constant} \dots (1) \text{ in which the units}$$

$$\text{are } \frac{m^2}{s^2} \text{ in each term}$$

THREE FORMS
OF BERNOULLI'S
EQUATION

If the fluid is incompressible ($\rho = \text{constant}$)

$$\frac{u^2}{2} + \frac{p}{\rho} + gz = \text{a constant} \dots (2) \left(\frac{m^2}{s^2} \text{ in each term}\right),$$

or $\dfrac{u^2}{2g} + \dfrac{p}{\rho g} + z = \text{a constant} \dots (3)$ (m in each term).

3.4 Relationship between Bernoulli's Equation and the Steady-flow Energy Equation

Steady-flow energy equation:

$$\frac{\dot{Q}_{\text{IN}}}{\dot{m}} - \frac{\dot{W}_{\text{OUT}}}{\dot{m}} + (e_1 + p_1 v_1 + \tfrac{1}{2}u_1^2 + gz_1) - (e_2 + p_2 v_2 + \tfrac{1}{2}u_2^2 + gz_2) = 0$$

in which each term has dimensions $\dfrac{m^2}{s^2}$.

33

Substituting $v = \dfrac{1}{\rho}$, $\dot{Q}_{IN} = -\dot{Q}_{OUT}$, $\dot{W}_{OUT} = -\dot{W}_{IN}$ and rearranging gives

$$\frac{p_1}{\rho_1} + \frac{1}{2}u_1^2 + gz_1 + \frac{\dot{W}_{IN}}{\dot{m}} = \frac{p_2}{\rho_2} + \frac{1}{2}u_2^2 + gz_2 + (e_2 - e_1) + \frac{\dot{Q}_{OUT}}{\dot{m}}.$$

Considering now an incompressible fluid, dividing throughout by g gives

$$\frac{p_1}{\rho g} + \frac{1}{2}\frac{u_1^2}{g} + z_1 + \frac{\dot{W}_{IN}}{\dot{m}g} = \frac{p_2}{\rho g} + \frac{1}{2}\frac{u_2^2}{g} + z_2 + \frac{(e_2 - e_1) + (\dot{Q}/\dot{m})}{g}. \quad \ldots (4)$$

Compare (4) with (3) allowing for a real fluid in which viscous forces are set up in opposition to accelerating forces. Work is needed to overcome these and mechanical energy is converted into internal energy and heat transfer. Also, energy may be added (e.g. from a pump) or removed (e.g. by a turbine).

i.e. $\dfrac{p_1}{\rho g} + \dfrac{1}{2}\dfrac{u_1^2}{g} + z_1 + H_T = \dfrac{p_2}{\rho g} + \dfrac{1}{2}\dfrac{u_2^2}{g} + z_2 + h_f$ (modified Bernoulli's equation).

$$\ldots (5)$$

The student is encouraged to study this comparison closely since the link between Bernoulli and the full steady-flow energy equation is an important one.

Comparing equations 4 and 5 we see that:

$$H_T \equiv \frac{\dot{W}_{IN}}{\dot{m}} \qquad \text{(the input head or work transfer);}$$

$$h_f \equiv \frac{(e_2 - e_1) + (\dot{Q}_{OUT}/\dot{m})}{g} \qquad \text{(the head 'lost' to viscous friction which increases internal energy and gives a heat transfer to surroundings).}$$

Thus equation 3 will derive directly from the steady-flow energy equation in the absence of any work transfer and any viscous friction.

Each term in equation 3 for Bernoulli has the dimensions of length and is given the name 'head' as follows:

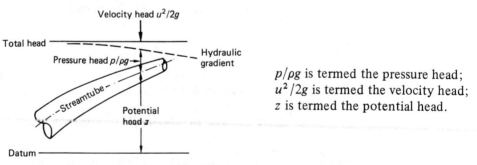

$p/\rho g$ is termed the pressure head; $u^2/2g$ is termed the velocity head; z is termed the potential head.

The hydraulic gradient represents the term $\left(\dfrac{p}{\rho g} + z\right)$, which is the static pressure measured by a manometer at the given point giving the manometric head.

3.5 Worked Examples

Example 3.1

A conical pipe varying in diameter from 1.37 m to 0.6 m forms part of a horizontal water main. The pressure head at the large end is found to be 30.5 m and at the small end 29.5 m. Find the discharge through the pipe neglecting all losses.

Answer

Bernoulli's equation between 1 and 2:

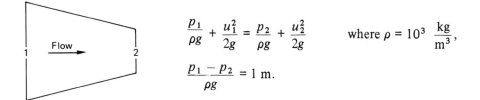

$$\frac{p_1}{\rho g} + \frac{u_1^2}{2g} = \frac{p_2}{\rho g} + \frac{u_2^2}{2g} \qquad \text{where } \rho = 10^3 \; \frac{\text{kg}}{\text{m}^3},$$

$$\frac{p_1 - p_2}{\rho g} = 1 \text{ m}.$$

Mass continuity (ρ constant):

$$A_1 u_1 = A_2 u_2,$$

$$u_2 = u_1 \frac{A_1}{A_2} = u_1 \left(\frac{D_1}{D_2}\right)^2 = u_1 \left(\frac{1.37}{0.6}\right)^2 = 5.214 u_1.$$

Substituting in Bernoulli: $\qquad 1 \text{ m} = \dfrac{u_1^2}{2g} (5.214^2 - 1) = 26.18 \; \dfrac{u_1^2}{2g}$

Thus $\qquad\qquad\qquad\qquad u_1 = \sqrt{\dfrac{2 \times 9.81 \; \dfrac{\text{m}}{\text{s}^2} \times 1 \text{ m}}{26.18}} = 0.866 \; \dfrac{\text{m}}{\text{s}}.$

And $\qquad\qquad\qquad\qquad \dot{V} = A_1 u_1 = \dfrac{\pi}{4} \times 1.37^2 \text{ m}^2 \times 0.866 \; \dfrac{\text{m}}{\text{s}}$

$$= 1.276 \; \frac{\text{m}^3}{\text{s}}.$$

Example 3.2

A conical tube, varying in diameter from 0.3 m to 0.15 m over a length of 1.5 m is fixed vertically with its smaller end upwards and forms part of a pipe line. Water flows down through the tube at a rate of 0.12 m³/s. The pressure at the upper end of the tube is equivalent to a head of 3 m water gauge. The energy degradation in the tube may be expressed as $0.3 \, (u_1 - u_2)^2/2g$ where u_1 and u_2 are the flow velocities at inlet and outlet of the tube, respectively. Determine the pressure head at the lower end of the tube.

Answer

Bernoulli from 1 to 2:

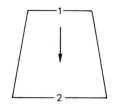

$$\frac{p_1}{\rho g} + z_1 + \frac{u_1^2}{2g} = \frac{p_2}{\rho g} + z_2 + \frac{u_2^2}{2g} + h_f$$

where h_f represents energy degradation expressed as a 'head'.

$$\dot{V} = 0.12 \ \frac{m^3}{s} \ .$$

Thus
$$u_1 = \frac{\dot{V}}{A_1} = \frac{\dot{V}}{\frac{1}{4}\pi D^2} = \frac{0.12 \ \frac{m}{s}}{\frac{1}{4}\pi \times 0.15^2 \ m^2} = 6.79 \ \frac{m}{s} \ .$$

Similarly,
$$u_2 = \frac{\dot{V}}{A_2} = \frac{0.12}{\frac{1}{4}\pi \times 0.3^2} = 1.698 \ \frac{m}{s} \ .$$

$$\frac{p_1}{\rho g} = 3 \ m; \qquad z_1 - z_2 = 1.5 \ m;$$

$$h_f = \frac{0.3}{2g} \ (u_1 - u_2)^2 = \frac{0.3}{2 \times 9.81 \ \frac{m}{s^2}} \ (6.79 - 1.698) \ \frac{m^2}{s^2} = 0.396 \ m.$$

Thus

$$\frac{p_2}{\rho g} = \left(3 + 1.5 + \frac{6.79}{2 \times 9.81} - \frac{1.698}{2 \times 9.81} - 0.396\right) m$$

$$= 6.307 \ m.$$

Example 3.3

A portion of a pipe for conveying water is vertical and varies in diameter from 25 mm at the lower end to 50 mm at the upper end over a length of 2 m. Pressure gauges are located at the inlet and outlet of this conical portion. When the quantity of water flowing up through the pipe is 0.003 m³/s the inlet gauge shows a pressure 30 kN/m² above that of the outlet gauge. Determine the quantity of water flowing through the pipe when the pressure gauges show zero pressure difference and the direction of flow is downwards. Assume that the energy degradation due to viscous friction may be expressed as $k\dot{V}^2$, where $\dot{V}$ is the quantity of water flowing and k is a constant.

Answer

Flow initially upwards:

Bernoulli:

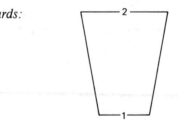

$$\frac{p_1}{\rho g} + \frac{u_1^2}{2g} + z_1 = \frac{p_2}{\rho g} + \frac{u_2^2}{2g} + z_2 + k\dot{V}^2$$

$$p_1 - p_2 = 30 \ \frac{kN}{m^2} \ ;$$

$$\frac{p_1 - p_2}{\rho g} = \frac{30 \ \frac{kN}{m^2}}{10^3 \ \frac{kg}{m^3} \times 9.81 \ \frac{m}{s^2}} \left[\frac{kg \ m}{N \ s^2}\right] = 3.058 \ m;$$

$$z_1 - z_2 = -2 \ m;$$

$$\dot{V} = 0.003 \ \frac{m^3}{s}; \qquad u_1 = \frac{\dot{V}}{A_1} = \frac{0.003 \ \frac{m^3}{s}}{\frac{1}{4}\pi \times 0.025 \ m^2} = 6.11 \ \frac{m}{s}.$$

Similarly
$$u_2 = \frac{0.003}{\frac{1}{4}\pi \times 0.05^2} = 1.53 \ \frac{m}{s}.$$

Thus
$$k\dot{V}^2 = \left(3.058 - 2 + \frac{6.11^2 - 1.53^2}{2 \times 9.81}\right) m = 2.841 \ m$$

and
$$k = 2.841 \times \left(\frac{1}{0.003}\right)^2 \frac{m \ s^2}{m^6} = 0.3157 \times 10^6 \ \frac{s^2}{m^5}.$$

Flow downwards: $p_1 = p_2$.

Thus
$$\frac{u_2^2}{2g} + z_2 = \frac{u_1^2}{2g} + z_1 + k\dot{V}^2.$$

$$z_2 - z_1 = \dot{V}^2 \left(k + \frac{1/A_1^2 - 1/A_2^2}{2g}\right) \qquad \left(\text{since } u = \frac{\dot{V}}{A}\right).$$

Now
$$A_1 = \tfrac{1}{4}\pi \times 0.025^2 = 0.49 \times 10^{-3} \ m^2,$$
$$A_2 = \tfrac{1}{4}\pi \times 0.05^2 = 1.96 \times 10^{-3} \ m^2 \qquad \text{and substituting,}$$

$$2 \ m = \dot{V}^2 \left(0.3157 \times 10 \ \frac{s^2}{m^5} + \frac{4.16 \times 10 - 0.26 \times 10^6 \ s^2}{2 \times 9.81 \ m \ m^4}\right)$$

$$= \dot{V}^2 \ (0.315 + 0.199) \times 10^6 = 0.514 \ \times 10^6 \ \dot{V}^2 \ \frac{s^2}{m^5}$$

or
$$\dot{V}^2 = 3.892 \times 10^{-6} \ \frac{m^6}{s^2}$$

or
$$\dot{V} = 1.973 \times 10^{-3} \ \frac{m^3}{s}.$$

Example 3.4

Oil of specific gravity 0.92 is being pumped at 0.0053 m³/s by a centrifugal pump from a supply tank to a hydrostatic journal bearing above the tank. Pressure gauges in the suction and discharge pipes read (-35) kN/m² and 550 kN/m² respectively when the vertical distance between the pressure tappings is 10 m. If the internal diameters of the suction and discharge pipes are 0.05 m and 0.076 m respectively, calculate the power supplied to the pump assuming an overall efficiency of the pump of 75 per cent.

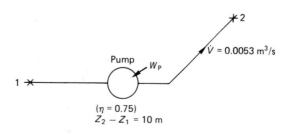

Answer

Bernoulli from 1 to 2:

$$\frac{p_1}{\rho g} + \frac{u_1^2}{2g} + z_1 + W_P = \frac{p_2}{\rho g} + \frac{u_2^2}{2g} + z_2$$

or $W_P = \dfrac{p_2 - p_1}{\rho g} + (z_2 - z_1) + \dfrac{u_2^2 - u_1^2}{2g}$

$$= \frac{\left[550 - (-35)\right]\frac{\text{kN}}{\text{m}^2}}{0.92 \times 10^3 \, \frac{\text{kg}}{\text{m}^3} \times 9.81 \, \frac{\text{m}}{\text{s}^2}} \left[\frac{\text{kg m}}{\text{N s}^2}\right] + 10 \text{ m}$$

$$+ \frac{0.0053^2 \, \frac{\text{m}^6}{\text{s}^2}}{2 \times 9.81 \, \frac{\text{m}}{\text{s}^2}} \left(\frac{4}{\pi}\right)^2 \left(\frac{1}{0.074^4 \, \text{m}^4} - \frac{1}{0.05^4 \, \text{m}^4}\right) \qquad \left(\text{since } u = \frac{\dot{V}}{A}\right)$$

$$= 64.82 \text{ m} + 10 \text{ m} + 0.294 \text{ m} = 75.11 \text{ m}.$$

$$\text{Power} = \frac{\rho \dot{V} (g W_P)}{\eta} = \frac{0.92 \times 10^3 \, \frac{\text{kg}}{\text{m}^3} \times 0.0053 \, \frac{\text{m}^3}{\text{s}} \times 9.81 \, \frac{\text{m}}{\text{s}^2} \times 75.11 \text{ m}}{0.75} \left[\frac{\text{N s}^2}{\text{kg m}}\right]$$

$$= 4790 \, \frac{\text{N m}}{\text{s}} = 4.79 \text{ kW}.$$

Example 3.5

A pump draws water from reservoir A and lifts it to reservoir B (see diagram). The internal diameters of the suction and discharge pipes are respectively 75 mm and 60 mm. Calculate the gauge pressures at the inlet and outlet of the pump and the power output of the pump when the discharge is 0.02 m³/s.

Assume that the energy degradation from A to pump inlet is given by $3u^2/2g$ where u is the flow velocity in the suction pipe and the energy degradation from pump outlet to B is given by $20u^2/2g$, where u is now the flow velocity in the discharge pipe.

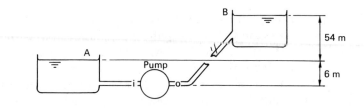

Answer

Bernoulli from A to pump inlet (i):

$$\frac{p_A}{\rho g} + \frac{u_A^2}{2g} + z_A = \frac{p_i}{\rho g} + \frac{u_i^2}{2g} + z_i + h_{f,A-i};$$

Bernoulli from pump outlet (o) to B:

$$\frac{p_o}{\rho g} + \frac{u_o^2}{2g} + z_o = \frac{p_B}{\rho g} + \frac{u_B^2}{2g} + z_B + h_{f,o-B};$$

$$z_A - z_i = 6 \text{ m}; \qquad \dot{V} = 0.02 \; \frac{\text{m}^3}{\text{s}}; \qquad z_B - z_o = 60 \text{ m}.$$

Thus $u_i = \dfrac{\dot{V}}{A_i} = \dfrac{0.02 \; \frac{\text{m}^3}{\text{s}}}{\frac{1}{4}\pi \times 0.075^2 \; \text{m}^2} = 4.527 \; \dfrac{\text{m}}{\text{s}}; \qquad \dfrac{u_i^2}{2g} = \dfrac{4.527^2}{2 \times 9.81} \text{ m} = 1.045 \text{ m};$

and $u_o = \dfrac{0.02}{\frac{1}{4}\pi \times 0.06^2} = 7.074 \; \dfrac{\text{m}}{\text{s}}; \qquad \dfrac{u_o^2}{2g} = \dfrac{7.074^2}{2 \times 9.81} \text{ m} = 2.55 \text{ m}.$

Thus $h_{f,A-i} = \dfrac{3 \times 4.527^2 \; \frac{\text{m}^2}{\text{s}^2}}{2 \times 9.81 \; \frac{\text{m}}{\text{s}^2}} = 3.134 \text{ m}; \qquad h_{f,o-B} = \dfrac{20 \times 7.074^2}{2 \times 9.81} = 51 \text{ m}.$

$$\frac{p_i}{\rho g} = z_A - z_i - \frac{u_i^2}{2g} - h_{f,A-i} = (6 - 1.045 - 3.134) \text{ m} = 1.821 \text{ m} \left(\frac{p_A}{\rho g} = \frac{u_A^2}{2g} = 0 \right)$$

$$\frac{p_o}{\rho g} = z_B - z_o + h_{f,o-B} - \frac{u_o^2}{2g} = (60 + 51 - 2.55) \text{ m} = 108.45 \text{ m};$$

$$p_i = 10^3 \; \frac{\text{kg}}{\text{m}^3} \times 9.81 \; \frac{\text{m}}{\text{s}^2} \times 1.821 \text{ m} \left[\frac{\text{N s}^2}{\text{kg m}} \right] \left[\frac{\text{kN}}{10^3 \text{ N}} \right] = 17.864 \; \frac{\text{kN}}{\text{m}^2};$$

$$p_o = \frac{10^3 \times 9.81 \times 108.45}{10^3} = 1063.9 \; \frac{\text{kN}}{\text{m}^2};$$

$$\text{Power} = \rho \dot{V} g \left(\frac{p_o}{\rho g} + \frac{u_o^2}{2g} - \frac{p_i}{\rho g} - \frac{u_i^2}{2g} \right)$$

$$= 10^3 \; \frac{\text{kg}}{\text{m}^3} \times 0.02 \; \frac{\text{m}^3}{\text{s}} \times \frac{9.81 \text{ m}}{\text{s}^2} \left[108.45 + 2.55 - 1.821 - 1.045 \right] \text{m} \left[\frac{\text{N s}^2}{\text{kg m}} \right] \left[\frac{\text{kN}}{10^3 \text{ N}} \right] \left[\frac{\text{kW}}{\text{kN m}} \right]$$

$$= 21.22 \text{ kW}.$$

Exercises

1 A pump draws water from a sump and discharges it into a tank in which the level of water is 80 m above that in the sump (diagram below). The internal diameters of the suction and discharge pipe are 100 mm and 50 mm respectively. The inlet and exit sections of the pump are in the same horizontal plane 6 m above the water level in the sump.

The energy degradation due to viscous friction in the suction pipe is twice the velocity head in that pipe; the corresponding energy degradation in the discharge is twenty-five times the velocity head in that pipe and the discharge pipe exit degradation is equal to the velocity head in that pipe.

When the power transmitted to the water by the pump is 40 kW, the pressure at inlet to the pump is (-7 m) water gauge. For these conditions calculate the flow

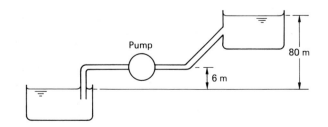

rate through the pump and the energy degradation between the pump inlet and outlet sections. (*Answer:* 0.0186 m³/s, 20.17 m)

2 In the system shown in the diagram below, water is raised through a vertical height of 4 m by a pump and discharged to atmosphere at the rate of 0.018 m³/s. The energy degradation due to friction in the pipe is 0.025 m of water per metre length on the suction side of the pump, and 0.05 m water per metre length on the discharge side.

Plot a curve showing the variation in pressure head through the system.

If the pump efficiency is 75 per cent, determine the power required to drive the pump. (*Answer:* 2.071 kW power)

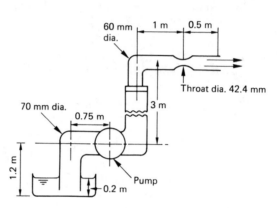

4 Fluid Measurement

4.1 Introduction

A full treatment of flow measurement would embrace the measurement of viscosity and pressure as well as the measurement of velocity and flow rate.

The work of this chapter is confined to meters for flow measurement based upon an induced pressure difference.

4.2 Venturimeter

In its simplest form the venturimeter is a short length of pipe tapering to a throat of smaller diameter in the middle. As the fluid passes through the venturi its dynamic or kinetic head ($u^2/2g$) increases as the throat is approached. Consequently, since the total head is unchanged in isenergic flow, the manometric head is reduced. A manometer is used to measure the manometric head between inlet and throat.

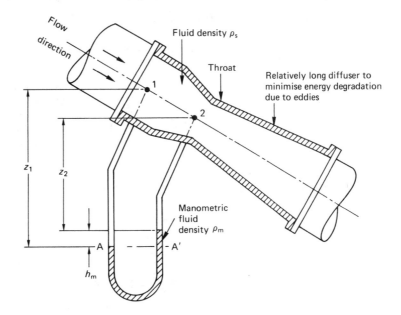

For an ideal fluid, from Bernoulli's equation:

$$\frac{p_1}{\rho_s g} + \frac{u_1^2}{2g} + z_1 = \frac{p_2}{\rho_s g} + \frac{u_2^2}{2g} + z_2, \qquad \text{and, rearranging,}$$

$$\frac{u_2^2 - u_1^2}{2g} = \left(\frac{p_1}{\rho_s g} + z_1\right) - \left(\frac{p_2}{\rho_s g} + z_2\right) = h_m \qquad \text{(the manometric head)}.$$

41

Continuity: $\qquad A_1 u_1 = A_2 u_2 \qquad$ (ρ_s constant)

$$u_1 = u_2 \frac{A_2}{A_1}, \qquad \text{and, substituting in the above,}$$

$$\frac{u_2^2 - u_1^2}{2g} = \frac{u_2^2}{2g}\left[1 - \left(\frac{A_2}{A_1}\right)^2\right] = h_m, \qquad \text{giving}$$

$$u_2 = \sqrt{2gh_m} \Big/ \sqrt{1 - \left(\frac{A_2}{A_1}\right)^2}.$$

and

$$\dot{V} = A_2 u_2 = A_2 \sqrt{2gh_m} \Big/ \sqrt{1 - \left(\frac{A_2}{A_1}\right)^2}.$$

For a horizontal venturimeter, $\qquad z_1 = z_2$

and thus

$$h_m = \frac{p_1 - p_2}{\rho_s g},$$

whence

$$\dot{V} = \frac{A_2 \sqrt{2\left(\dfrac{p_1 - p_2}{\rho_s}\right)}}{\sqrt{1 - \left(\dfrac{A_2}{A_1}\right)^2}}.$$

Applying manometric principles to common level A–A' in the diagram (p. 41):

$$p_A = p_1 + \rho_s g z_1; \qquad p_{A'} = p_2 + \rho_s g (z_2 - h) + \rho_m g h;$$

But $\qquad p_A = p_{A'} \qquad$ (same level) and thus

$$p_1 + \rho_s g z_1 = p_2 + \rho_s g (z_2 - h) + \rho_m g h.$$

Dividing by $\rho_s g$ and rearranging, we get

$$\left(\frac{p_1}{\rho_s g} + z_1\right) - \left(\frac{p_2}{\rho_s g} + z_2\right) = h\left(\frac{\rho_m}{\rho_s} - 1\right) = h_m.$$

Hence

$$\dot{V} = \frac{A_2 \sqrt{2gh\left(\dfrac{\rho_m}{\rho_s} - 1\right)}}{\sqrt{1 - \left(\dfrac{A_2}{A_1}\right)^2}} = \frac{A_2 \sqrt{2gh\left(\dfrac{S_m}{S_s} - 1\right)}}{\sqrt{1 - \left(\dfrac{A_2}{A_1}\right)^2}}$$

where S is specific gravity.

Real Fluid

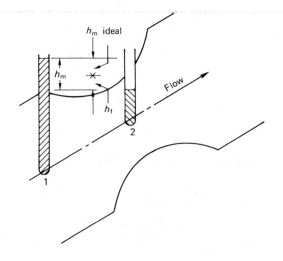

Viscous friction between the venturi inlet and throat causes a loss of head or energy degradation h_f. The manometric head measured will be greater than that for an ideal fluid and both the velocity and discharge based on the ideal equations will be too large. Thus the velocity and discharge equations must be corrected.

Defining, $\quad C_u = \dfrac{\text{actual velocity}}{\text{ideal velocity}} \quad$ (the coefficient of velocity),

$$u_2 = \frac{C_u \sqrt{2gh_m}}{\sqrt{1 - \left(\dfrac{A_2}{A_1}\right)^2}} \qquad \ldots (a)$$

Defining, $\quad C_D = \dfrac{\text{actual discharge}}{\text{ideal discharge}} \quad$ (the coefficient of discharge),

$$\dot{V} = \frac{A_2 C_D \sqrt{2gh_m}}{\sqrt{1 - \left(\dfrac{A_2}{A_1}\right)^2}} .$$

In this particular application, $C_D = C_u$. Since the measured manometric head is greater than the ideal manometric head, both C_D and C_u are less than unity. (Usually they are about 0.97.)

Applying Bernoulli's equation, for a real fluid:

$$\frac{p_1}{\rho_s g} + \frac{u_1^2}{2g} + z_1 = \frac{p_2}{\rho_s g} + \frac{u_2^2}{2g} + z_2 + h_m .$$

Hence $\quad \dfrac{u_2^2 - u_1^2}{2g} = \left(\dfrac{p_1}{\rho_s g} + z_1\right) - \left(\dfrac{p_2}{\rho_s g} + z_2\right) - h_f = h_m - h_f$

and

$$u_2 = \frac{\sqrt{2g\,(h_m - h_f)}}{\sqrt{1 - \left(\dfrac{A_2}{A_1}\right)^2}} . \qquad \ldots (b)$$

Comparing (a) and (b), we see that

$$C_u \sqrt{h_m} = \sqrt{h_m - h_f},$$

from which $\qquad h_f = (1 - C_u)^2\, h_m .$

4.3 Sharp-edged Orifice

This is a plate with a hole drilled in it which may be chamfered to a sharp edge. It may be fixed into the side of a tank or into a pipe or duct, and is a cheaper type of flow measurement device than the venturimeter, although it involves larger values of energy degradation.

After passing through the orifice the fluid stream continues to contract for a short distance downstream (approximately equal to the orifice diameter) before becoming parallel-sided. At this section the pressure across the fluid stream is uniform and the cross-sectional area of the stream is a minimum known as the vena contracta.

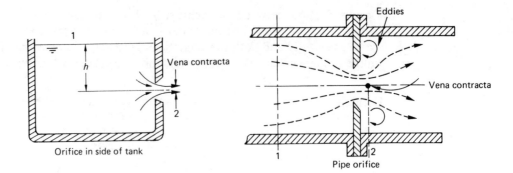

Eddies

Vena contracta

Orifice in side of tank

Vena contracta

Pipe orifice

In the pipe orifice case the fluid stream expands beyond the vena contracta to fill the pipe once more, but this is accompanied by the formation of eddies and a considerable energy degradation.

Orifice in the Side of a Tank

Since the upstream area is large compared with the areas of the orifice and the vena contracta, the upstream velocity, known as the approach velocity, can be taken as zero.

Bernoulli's equation from water surface to vena contracta:

$$h = \frac{u_2^2}{2g} \qquad \text{or} \qquad u_2 = \sqrt{2gh} \qquad \text{(ideal fluid)}.$$

For a real fluid, allowing for energy degradation:

$$u_2 = C_u \sqrt{2gh}$$

and thus

$$\dot{V} = A_2 u_2 = A_2 C_u \sqrt{2gh} .$$

It is inconvenient to include the unknown area of the vena contracta, A_2.

Defining, $\quad C_c = \dfrac{A_2}{A_O} = \dfrac{\text{area of vena contracta}}{\text{orifice area}} \qquad$ (the coefficient of contraction),

or

$$A_2 = A_O C_c ,$$

whence $\quad \dot{V} = A_O C_c C_u \sqrt{2gh} \qquad$ (both C_c and C_u are obtained experimentally).

Now for an ideal fluid (no energy degradation, no contraction),

$$\dot{V} = A_O \sqrt{2gh},$$

and since the coefficient of discharge $C_D = \dfrac{\text{actual discharge}}{\text{ideal discharge}}$,

then

$$C_D = \frac{A_O C_c C_u \sqrt{2gh}}{A_O \sqrt{2gh}} = C_c C_u$$

and

$$\dot{V} = A_O C_D \sqrt{2gh}.$$

The Pipe Orifice

Applying Bernoulli's equation between an upstream section (1) and the vena contracta (2) gives a result similar to that of the venturimeter.

$$u_2 = \frac{C_u \sqrt{2gh_m}}{\sqrt{1 - \left(\frac{A_2}{A_1}\right)^2}}$$

and

$$\dot{V} = A_2 u_2 = \frac{A_2 C_u \sqrt{2gh_m}}{\sqrt{1 - \left(\frac{A_2}{A_1}\right)^2}}.$$

replacing A_2 by $A_O C_C$ gives

$$\dot{V} = \frac{A_O C_C C_u \sqrt{2gh_m}}{\sqrt{1 - \left(C_C \frac{A_O}{A_1}\right)^2}} = \frac{A_O C_D \sqrt{2gh_m}}{\sqrt{1 - C_C^2 \left(\frac{A_O}{A_1}\right)^2}}.$$

Because this expression is complex, involving two coefficients, the flow equation is often reduced to either:

$$\dot{V} = \frac{A_O C_D \sqrt{2gh_m}}{\sqrt{1 - \left(\frac{A_O}{A_1}\right)^2}} \qquad \text{(similar to a venturimeter),}$$

or

$$= A_O C \sqrt{2gh_m} \qquad \text{(similar to the tank orifice where } C \text{ is not the coefficient of discharge).}$$

4.4 The Flow Nozzle

This is essentially a venturimeter with the diffuser omitted. It can be fitted within or at the end of a pipeline. The expressions obtained for the discharge through a venturimeter will apply to a nozzle. The absence of the diffuser will cause the formation of eddies and energy degradation in a nozzle will be much higher than in a venturimeter.

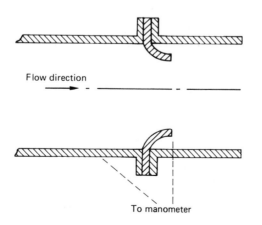

Flow direction

To manometer

4.5　The Pitot Tube

In its simplest form a pitot tube is a glass tube with its end bent through 90° placed in a stream of liquid with its opening facing upstream. Initially, liquid flows into the tube until its kinetic energy has been converted into potential energy. The liquid-filled tube then behaves like a solid body.

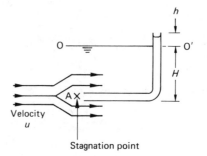

The liquid stream flows around the tube and the central streamline with an energy content $(p/\rho g + u^2/2g)$ (where p and u are the undisturbed values upstream of the tube) and is brought to rest at the nose of the pitot tube, point A in the diagram, without degradation of energy. Therefore the pressure head at A is $(p/\rho g + u^2/2g)$. But the pressure head at A is equal to the static head of liquid in the tube, i.e. $(h + H)$.

For the case with a free liquid surface at OO', $\dfrac{p}{\rho g} = H$ (diagram).

Hence
$$h = \frac{u^2}{2g}$$

or
$$u = \sqrt{2gh}.$$

Because the liquid at A is stagnant, A is called a stagnation point and the pressure there is termed the stagnation pressure.

4.6　The Pitotstatic Tube

In the case of OO' in the diagram above not being a free surface, $p/\rho g \neq H$ and some means of measuring $p/\rho g$ is required.

Pitotstatic tubes are a combination of pitot and static tubes; they can be separate or combined in the same sheath (diagram below).

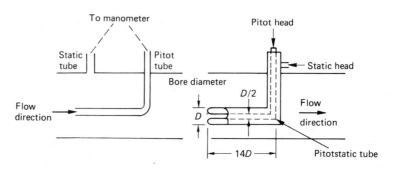

The pitot tube will measure the pitot head $(= p/\rho g + u^2/2g)$.

The static tube will measure the static head $(= p/\rho g)$.

By connecting across a manometer, the manometric head h_m will equal the difference, namely $u^2/2g$, giving

$$h_m = u^2/2g \qquad \text{or} \qquad u = \sqrt{2gh_m}.$$

The pitotstatic tube combines these two tubes and will measure only the local velocity of flow; it will not measure flow rate unless a traverse over the duct is carried out. If well designed according to British Standards there is no need for a meter correction coefficient; if not, $u = C\sqrt{2gh_m}$.

Values of coefficient of discharge for all common fluid-flow configurations can be obtained by reference to British Standard 1042.

4.7 Worked Examples

Example 4.1

Oil with a specific gravity of 0.88 is conveyed along a horizontal pipe line of 50 mm diameter, containing a venturimeter with a throat diameter of 0.4 times the pipe diameter. The pressure drop between entry and the throat is measured on a U-tube manometer, containing mercury of specific gravity 13.6, and the difference in levels of the columns is read as 'x' mm. The discharge coefficient for the venturimeter is 0.95. Show that the rate of discharge is

$$\dot{V} = 1.61 \times 10^{-4} \sqrt{x} \ \text{m}^3/\text{s}.$$

How will this expression be modified if the pipe line is inclined at θ to the horizontal?

Answer

$$\dot{V} = \frac{C_D A_2 \sqrt{2gh\left(\dfrac{S_m}{S_s} - 1\right)}}{\sqrt{1 - \left(\dfrac{A_2}{A_1}\right)^2}} = \frac{C_D A_1 A_2 \sqrt{2gh\,(K - 1)}}{\sqrt{A_1^2 - A_2^2}} \qquad \left(K = \frac{S_m}{S_s}\right),$$

which is independent of z term.

$$A_1 = \tfrac{1}{4}\pi\,(0.05)^2 \ \text{m}^2 = 19.63 \times 10^{-4} \ \text{m}^2 ; \qquad A_2 = (0.4)^2 A_1 = 3.14 \times 10^{-4} \ \text{m}^2 ;$$

$$V = \frac{0.95 \times 19.63 \times 3.14 \times 10^{-3} \ \text{m}^4}{\sqrt{(19.62^2 - 3.14^2) \times 10^{-3} \ \text{m}^4}} \sqrt{2 \times 9.81 \ \frac{\text{m}}{\text{s}} \ \frac{x}{1000} \ \text{m} \left[\frac{13.6}{0.88} - 1\right]}$$

$$= 3.022 \times 10^{-4} \ \text{m}^2 \times 0.532 \sqrt{x} \ \frac{\text{m}}{\text{s}} \qquad (x \text{ in mm})$$

$$= 1.609 \times 10^{-4} \sqrt{x} \ \frac{\text{m}^3}{\text{s}} .$$

The expression shows no change with change in θ.

Example 4.2

A venturimeter is used for measuring the flow of petrol up a pipe inclined at 35° to the horizontal. Pressure tappings are provided at the inlet and throat of the meter at a distance 0.45 m apart. The inlet and throat diameters are 0.3 m and 0.15 m respectively and the difference in mercury levels on a manometer connected across the meter is 50 mm. Calculate the flow in m^3/s. Take $C_D = 0.98$ for the meter.

If the manometer is now replaced by two Bourdon pressure gauges at inlet and throat, find the difference between the gauge readings for the same flow rate. Specific gravities are 0.78 for petrol and 13.6 for mercury.

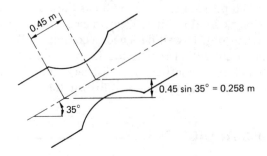

Answer

$$A_1 = \tfrac{1}{4}\pi \times 0.3^2 \ m^2 = 0.0707 \ m^2 \ ;$$

$$A_2 = \tfrac{1}{4}\pi \times 0.15^2 \ m^2 = 0.0177 \ m^2 \ .$$

As in example 4.1, with the original form for $\dot{V}$,

$$\dot{V} = \frac{0.98 \times 0.0707 \times 0.0177 \ m^4}{\sqrt{(0.0707^2 - 0.0177^2) \ m^4}} \ \sqrt{2 \times 9.81 \left(\frac{13.6}{0.78} - 1\right) 0.05 \ \frac{m^2}{s^2}}$$

$$= 0.0179 \ m^2 \times 4.01 \ \frac{m}{s}$$

$$= 0.072 \ \frac{m^3}{s} \ .$$

Now
$$\dot{V} = k\sqrt{h_m} \ ,$$

where
$$k = \frac{C_D A_1 A_2 \sqrt{2g}}{\sqrt{(A_1^2 - A_2^2)}}$$

and
$$h_m = \left(\frac{13.6}{0.78} - 1\right) \times 0.05,$$

i.e.
$$h_m = 0.82 \ m \ petrol.$$

But in manometry
$$h_m = \frac{p_1 - p_2}{\rho g} + z_1 - z_2 \ .$$

Thus
$$0.82 \ m = \frac{\Delta p}{780 \ \frac{kg}{m^3} \times 9.81 \ \frac{m}{s^2}} \ \left[\frac{kg \ m}{N \ s^2}\right] + 0 - 0.258 \ m,$$

$$\Delta p = 8248 \ m \ \frac{N}{m^3} = 8248 \ \frac{N}{m^2} \ .$$

Example 4.3

Petrol from a tank passes through a vertical venturimeter having an inlet diameter of 150 mm and a throat diameter of 37.5 mm. The inlet is 380 mm above the throat. Pressure gauges are fitted at inlet and throat. Assuming that energy degradation between inlet and throat may be expressed as $35\,u^2/2g$, u being the flow velocity at inlet, the pressure difference between inlet and throat is found to be $31\ kN/m^2$. Find the coefficient of discharge and the flow through the meter in m^3/s. Assume that the specific gravity for the petrol is 0.78.

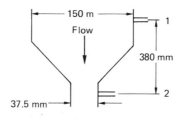

Bernoulli from 1 to 2:

$$\frac{p_1}{\rho g} + \frac{u_1^2}{2g} + z_1 = \frac{p_2}{\rho g} + \frac{u_2^2}{2g} + z_2 + h_f \qquad \text{where } h_f = 35\,\frac{u_1^2}{2g}.$$

Continuity from 1 to 2:

$$A_1 u_1 = A_2 u_2 \qquad (\rho \text{ constant}).$$

Thus

$$\frac{p_1 - p_2}{g} + z_1 - z_2 = \frac{u_2^2}{2g}\left[1 + 35\left(\frac{A_2}{A_1}\right)^2 - \left(\frac{A_2}{A_1}\right)^2\right]$$

$$h = \frac{u_2^2}{2g}\left[1 + 34\left(\frac{1}{16}\right)^2\right]$$

$$= 1.1328\,\frac{u_2^2}{2g}.$$

Thus

$$u_2 = \sqrt{\frac{1}{1.1328}}\,\sqrt{2g\,(h)}$$

where $h = \dfrac{p_1 - p_2}{\rho g} + z_1 - z_2$ (the manometric head)

and

$$\dot V = A_2 u_2 = 0.94 A_2 \sqrt{2g\,(h)}.$$

For no energy degradation $u_{2\,\text{ideal}} = \sqrt{2g\,(h)}\,\sqrt{\left[\dfrac{1}{1 - \dfrac{1}{16^2}}\right]}$

and

$$\dot V_{\text{ideal}} = A_2 u_{2\,\text{ideal}} = 1.002 A_2 \sqrt{2g\,(h)}.$$

Thus $C_D = \dfrac{\dot V}{\dot V_{\text{ideal}}} = \dfrac{0.94}{1.002} = 0.938.$

Manometric head $h_m = \dfrac{p_1 - p_2}{\rho g} + z_1 - z_2$

$$= 31\,000\ \frac{N}{m^2}\ \frac{m^3}{780\ kg}\ \frac{s^2}{9.81\ m}\ \left[\frac{kg\ m}{N\ s^2}\right] + 0.38\ m$$

$$= 4.431\ m\ petrol.$$

$$\dot{V} = C_{\mathrm{D}} A_2 \sqrt{2gh}$$

$$= 0.938 \times \tfrac{1}{4}\pi \, (0.0375)^2 \ \mathrm{m^2} \sqrt{2 \times 9.81 \ \tfrac{\mathrm{m}}{\mathrm{s^2}} \times 4.431 \ \mathrm{m}}$$

$$= 9.66 \times 10^{-3} \ \tfrac{\mathrm{m^3}}{\mathrm{s}} .$$

Example 4.4

The venturi in the diagram is being used as a suction device. Working from Bernoulli's equation, determine the volumetric flow rate of air through the venturi at the instant when suction of the liquid into the air stream commences. Assume that air in this instance is an ideal incompressible fluid. Take the densities of air and liquid to be 1.25 kg/m³ and 800 kg/m³ respectively.

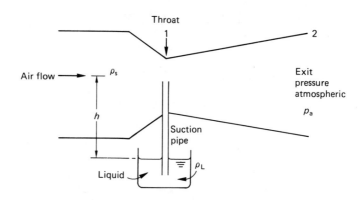

Answer

Data:
$$A_1 = 2 \ \mathrm{cm^2}; \qquad A_2 = 8 \ \mathrm{cm^2};$$
$$h = 0.25 \ \mathrm{m}; \qquad d = 0.15 \ \mathrm{m}.$$

Bernoulli from 1 to 2:

$$\frac{p_1}{\rho g} + \frac{u_1^2}{2g} = \frac{p_2}{\rho g} + \frac{u_2^2}{2g} \qquad \text{where } p_2 = p_{\mathrm{a}}.$$

Continuity equation from (1) to (2):

$$\rho A_1 u_1 = \rho A_2 u_2$$
or
$$A_1 u_1 = A_2 u_2 .$$
Thus
$$u_1 = \tfrac{8}{2} u_2 = 4 u_2 .$$
Thus combining the two equations,

$$\left. \begin{array}{l} \dfrac{p_{\mathrm{a}} - p_1}{\rho g} = 15 \ \dfrac{u_2^2}{2g}; \\[2ex] \text{For suction tube,} \qquad p_{\mathrm{a}} - p_1 = \rho_{\mathrm{L}} g \, h. \end{array} \right\}$$

Thus
$$u_2 = \sqrt{\frac{\rho_L}{\rho_s} \frac{2g}{15}} \; h = \sqrt{\frac{2}{15} \times \frac{800}{1.25} \times 0.25 \text{ m} \times 9.81 \frac{\text{m}}{\text{s}^2}}$$

$$= 14.46 \frac{\text{m}}{\text{s}} .$$

$$\dot{V} = A_2 u_2 = 8 \text{ cm}^2 \left[\frac{\text{m}^2}{10^4 \text{ cm}^2} \right] \times 14.46 \frac{\text{m}}{\text{s}} = 0.011\,57 \frac{\text{m}^3}{\text{s}} .$$

Example 4.5

An orifice of diameter 0.025 m discharges water at the rate of 13.25 litres per second when the head above the orifice is 61 m. If the jet diameter is 0.023 m determine:
(a) the value of all the coefficients,
(b) the power in the jet,
(c) the energy degradation in the orifice.

Answer

$$\text{Ideal velocity} = \sqrt{2gh} = \sqrt{2 \times 9.81 \frac{\text{m}}{\text{s}^2} \times 61 \text{ m}} = 34.6 \frac{\text{m}}{\text{s}} .$$

$$\text{Actual velocity} = \frac{\dot{V}}{A} = \frac{13.25 \frac{1}{\text{s}} \left[\frac{\text{m}^3}{10^3 \text{ l}} \right]}{\frac{1}{4}\pi \times 0.023^2 \text{ m}^2} = 31.89 \frac{\text{m}}{\text{s}} .$$

Thus
$$\text{coefficient of velocity } C_u = \frac{31.89}{34.6} = 0.922.$$

$$\text{Coefficient of contraction } C_C = \frac{\text{jet area}}{\text{orifice area}} = \left(\frac{0.023}{0.025} \right)^2 = 0.846.$$

$$\text{Coefficient of discharge} = C_D = C_u \times C_C = 0.78.$$

$$\text{Jet power} = \rho \dot{V} \frac{u^2}{2} = \frac{\rho \dot{V}^3}{2A^2} = \frac{13.25^3 \left(\frac{1}{\text{s}} \right)^3 \left[\frac{\text{m}^9}{10^9 \text{ l}^3} \right] 10^3 \frac{\text{kg}}{\text{m}^3}}{2 \left(\frac{1}{4}\pi \times 0.023^2 \right)^2 \text{ m}^4} .$$

$$= 6738 \frac{\text{kg m}^2}{\text{s}^3} \left[\frac{\text{N s}^2}{\text{kg m}} \right] \left[\frac{\text{kW s}}{10^3 \text{ N m}} \right] = 6.738 \text{ kW}.$$

$$h_f = h (1 - C_u^2) = 61 \text{ m} (1 - 0.922^2) = 9.14 \text{ m}.$$

Example 4.6

An orifice of diameter 0.025 m is situated in a plate which divides a tank containing water into two compartments. On one side of the plate, the water level is 6 m above the orifice and the water surface is at atmospheric pressure; on the other side, the water level is 1.2 m above the orifice and the air above the water is held at a pressure of 70 kN/m² gauge. Determine the water flow rate through the orifice if the coefficient of discharge is 0.80.

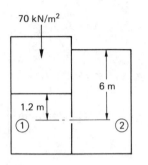

70 kN/m²

6 m

1.2 m

① ②

Answer

$$p_1 = p_0 + \rho g h_1$$

$$= 70 \ \frac{kN}{m^2} + \left(10^3 \ \frac{kg}{m^3} \times 9.81 \ \frac{m}{s^2} \times 1.2 \ m\right) \left[\frac{N \ s^2}{kg \ m}\right] \left[\frac{kN}{10^3 \ N}\right]$$

$$= 81.77 \ \frac{kN}{m^2} \ .$$

$$p_2 = p_0 + \rho g h = 0 + \left(\frac{10^3 \times 9.81 \times 6}{10^3}\right) = 58.86 \ \frac{kN}{m^2} \ .$$

Thus flow direction is from 1 to 2 and

$$p_1 - p_2 = 22.91 \ \frac{kN}{m^2} = \Delta p,$$

whence

$$\dot{V} \ A_0 \ C_D \ \sqrt{\frac{2 \Delta p}{\rho}} = 0.8 \times \tfrac{1}{4} \pi \times 0.025^2 \ m^2 \ \sqrt{\frac{2 \times 22.91 \ \dfrac{kN}{m^2}}{10^3 \ \dfrac{kg}{m^3}} \left[\frac{kg \ m}{N \ s^2}\right]} \ ,$$

$$\dot{V} = 2.658 \times 10^{-3} \ \frac{m^3}{s} \ .$$

Example 4.7

An orifice of diameter 0.030 m measures the rate of flow of air along a pipe of diameter 0.050 m. The coefficient of discharge for the orifice is 0.063 and its coefficient of contraction is 0.63. Show, from first principles, that if H is the difference of pressure head between upstream in the orifice and the vena contracta in metres of water, and ρ is the density of air in kg/m³, the rate of flow in standard m³/min is given by $\dot{V}_s = 0.672\sqrt{2g\rho H}$, the mass of the standard m³ of air being 1.24 kg.

Answer

From first principles (see section 4.3),

$$\dot{V} = \frac{A_2 C_D \sqrt{2gh}}{\sqrt{1 - C_C^2 \left(\dfrac{A_2}{A_1}\right)^2}}$$

52

where h is head across the orifice.

Assuming a horizontal pipe, $\quad h = H \dfrac{\rho_w}{\rho_{air}} = H \dfrac{10^3}{\rho}$.

Thus $\quad \dot{V} = \dfrac{\frac{1}{4} \pi \times 0.03^2 \text{ m}^2 \times 0.603 \sqrt{2 \times g \dfrac{\text{m}}{\text{s}^2} \times H \,(10^3/\rho)\, \text{m}}}{\sqrt{1 - 0.63^2 \left(\dfrac{0.03}{0.05}\right)^4}}$, $\quad$ with ρ in $\dfrac{\text{kg}}{\text{m}^3}$.

$$= 0.0138 \sqrt{\dfrac{2Hg}{\rho}} \dfrac{\text{m}^3}{\text{s}} .$$

But from continuity $\qquad\qquad \rho_s \dot{V}_s = \rho \dot{V}.$

Thus $\qquad \dot{V}_s = \dfrac{\rho \dot{V}}{\rho_s} = \dfrac{\rho}{1.24 \dfrac{\text{kg}}{\text{m}^3}} \times 0.0138 \sqrt{\dfrac{H}{\rho}} \dfrac{\text{m}^3}{\text{s}} \left[\dfrac{60\text{ s}}{\text{min}}\right]$

$$= 0.67 \sqrt{2g\rho H} \dfrac{\text{m}^3}{\text{min}} .$$

Example 4.8

A pitotstatic tube was placed in the centre of a pipe of diameter 0.2 m diameter with one outlet facing the stream and the other perpendicular to it. The difference of pressure between the two outlets was 0.037 m of water. The coefficient of the tube was unity.

Taking the mean velocity of the water in the pipe to be 0.83 times the maximum velocity, find the discharge through the pipe.

Answer

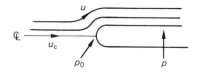

Area of pipe $= A = \frac{1}{4} \pi \, (D)^2 = \frac{1}{4} \pi \, (0.2)^2 \text{ m}^2 = 0.0314 \text{ m}^2$.

From the Bernoulli equation for the pipe centre-line,

$$u_C = \sqrt{2g \dfrac{\Delta p}{\rho g}} ,$$

where $\Delta p = p_0 - p$ and p_0 is the stagnation pressure.

$$u_C = \sqrt{2 \times 9.81 \dfrac{\text{m}}{\text{s}^2} \, 0.037 \text{ m}} = 0.852 \dfrac{\text{m}}{\text{s}} \text{ for water.}$$

Thus $\qquad u_m = 0.83 \, u_C = 0.707 \dfrac{\text{m}}{\text{s}} \qquad$ (mean velocity),

and $\qquad \dot{V} = A u_m = 0.0314 \text{ m}^2 \times 0.707 \dfrac{\text{m}}{\text{s}}$

$$= 0.0222 \dfrac{\text{m}^3}{\text{s}} .$$

Exercises

1 Derive an expression for the volume flow rate through a venturimeter and explain the effect of viscous friction on the measurement of head and how this may be allowed for.

A pipe of diameter 100 mm has to carry a maximum flow rate of water of 0.0341 m^3/s which is to be measured by means of a horizontally mounted venturimeter and a mercury U-tube manometer whose maximum scale length is 600 mm. Determine the throat diameter of the venturimeter which will be required to give a maximum manometer deflection at the maximum flow. Assume that C_D is 0.96 for the meter and that the specific gravity of mercury is 13.6. (*Answer:* 0.0591 m)

2 Water flows steadily through a horizontal venturimeter which has a throat-to-inlet diameter ratio of 0.5. A U-tube mercury manometer connected between an upstream tapping and a tapping at the throat registers a difference of 250 mm between the mercury levels. Working from Bernoulli's equation, determine the velocity of the water at the throat. Ignore all energy degradation.

(*Answer:* 8.12 m/s)

3 The centre of an orifice is situated 0.152 m from the bottom of a vessel containing water to a depth of 0.762 m. Water issues from the orifice horizontally and strikes the horizontal plane through the base of the vessel at a distance of 0.597 m from the vena contracta. Determine:
(a) the value of the coefficient of velocity of the orifice,
(b) the energy degradation in the orifice in m water.
(*Hint:* The issuing jet declines vertically under a constant gravitational acceleration. Use the laws of mechanics.) (*Answer:* 0.98, 0.0238 m)

4 A jet of water issues from an orifice under a head h. Show that the angle of impact of the water striking a horizontal table situated at a vertical distance x below the orifice is given by

$$C_u^2 \tan^2 \theta = \frac{x}{h}.$$

A vessel resting on a horizontal table has a convergent–divergent nozzle 0.20 m above the base. The vessel is filled with water to a height of 1.40 m above the base. The nozzle is of diameter 0.05 m and it runs full. If the jet strikes the table at an angle of 23° to the horizontal, calculate:
(a) the coefficient of velocity of the nozzle,
(b) the discharge rate.
(Assume the coefficient of contraction for the nozzle is unity.)
(*Answer:* 0.961, 0.009 16 (m^3/s))

5 The Momentum Equation for the Steady Flow of an Inviscid Incompressible Fluid

5.1 Introduction

Newton's second law of motion:

$$\text{force} = \text{rate of change of momentum},$$

or

$$F = \frac{\mathrm{d}}{\mathrm{d}t}(mu).$$

This strictly is a vector equation with components in any three perpendicular directions, say x, y and z.

Thus for a system in which several forces are acting:

resolving in the x-direction, $\Sigma F_x = \dfrac{\mathrm{d}}{\mathrm{d}t}(mu)_x$;

resolving in the y-direction, $\Sigma F_y = \dfrac{\mathrm{d}}{\mathrm{d}t}(mu)_y$;

resolving in the z-direction, $\Sigma F_z = \dfrac{\mathrm{d}}{\mathrm{d}t}(mu)_z$.

where x, y and z are the components in each direction.

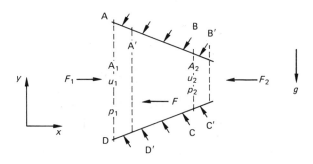

Consider the steady one-dimensional flow of a fluid within a stream tube in the x-direction. For a section ABCD (called a control volume) of the stream tube the forces exerted at the boundary will be:

F_1 on end AD due to pressure p_1;
F_2 on end BC due to pressure p_2;
F, the resultant of the pressure forces between AD and BC, also acts in the x-direction.

Thus $\Sigma F = F_1 - F_2 - F$ (the weight force is not included because there is no component in the x-direction).

During a small interval of time δt the fluid contained within ABCD will move to A'B'C'D'.

But for steady conditions the momentum within ABCD will remain constant. Thus any change of momentum will be equal to the difference between the momentum leaving at section 2 and that entering at section 1.

For mass flow rate $\dot{m}$, the momentum entering at 1 = $(\dot{m}\,\delta t) \times u_1$.
For mass flow rate $\dot{m}$, the momentum leaving at 2 = $(\dot{m}\,\delta t) \times u_2$.
Change of momentum = $(\dot{m}\,\delta t) \times (u_2 - u_1)$.
Change in momentum flux = $\dot{m}(u_2 - u_1)$.
Hence by Newton's law, $\Sigma F = F_1 - F_2 - F = \dot{m}(u_2 - u_1)$ for the control volume ABCD.

Strictly for the stream tube shown, it is clear that there must be velocity components in the y-direction. Thus u_1 and u_2 are the velocity components in the x-direction.

Writing
$$\left.\begin{array}{l} F_1 = p_1 A_1 \\ F_2 = p_2 A_2 \end{array}\right\} \text{ we get:}$$

$$p_1 A_1 - p_2 A_2 - F = \dot{m}(u_2 - u_1).$$

Writing $\quad \dot{m} = \rho \dot{V} = \rho(A_1 u_1) - \rho(A_2 u_2),$

thus $\quad p_1 A_1 - p_2 A_2 - F = \rho \dot{V}(u_2 - u_1) = \rho(A_2 u_2^2 - A_1 u_1^2).$

Similarly for a two-dimensional stream tube:

$$\Sigma F_x = F_1 \cos\theta_1 - F_2 \cos\theta_2 - F \cos\alpha,$$

$$\Sigma F_y = F_1 \sin\theta_1 - F_2 \sin\theta_2 + F \sin\alpha - W.$$

$$\frac{d}{dt}(mu)_x = \dot{m}(u_2 \cos\theta_2 - u_1 \cos\theta_1),$$

$$\frac{d}{dt}(mu)_y = \dot{m}(u_2 \sin\theta_2 - u_1 \sin\theta_1).$$

Newton's second law becomes:

$$p_1 A_1 \cos \theta_1 - p_2 A_2 \cos \theta_2 - F \cos \alpha = \dot{m}(u_2 \cos \theta_2 - u_1 \cos \theta_1)$$
$$= \rho \dot{V}(u_2 \cos \theta_2 - u_1 \cos \theta_1)$$
$$= \rho (A_2 u_2^2 \cos \theta - A_1 u_1^2 \cos \theta).$$

$$p_1 A_1 \sin \theta_1 - p_2 A_2 \sin \theta_2 + F \sin \alpha - W = \dot{m}(u_2 \sin \theta_2 - u_1 \sin \theta_1)$$
$$= \rho \dot{V}(u_2 \sin \theta_2 - u_1 \sin \theta_1)$$
$$= \rho (A_2 u_2^2 \sin \theta_2 - A_1 u_1^2 \sin \theta_1)$$

from which F and α can be obtained.

5.2 Applications

Impact of a Jet Onto a Stationary Plate

Horizontal jet, plane surface perpendicular to oncoming fluid after striking the surface of the jet

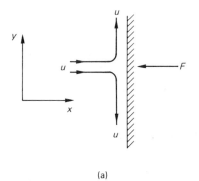

(a)

$$\Sigma F_x = -F = \dot{m}(0 - u) = -\dot{m}u,$$
$$F = \dot{m}u = \rho A u^2$$

Vertical jet, plane surface perpendicular to oncoming fluid

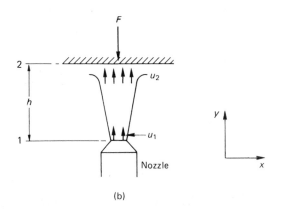

(b)

The jet velocity at impact will be less than u_1. Apply Bernoulli's equation between 1 and 2 to give:

$$\frac{u_2^2}{2g} = \frac{u_1^2}{2g} - h \qquad \text{and hence } u_2.$$

$$\Sigma F_y = -F = \dot{m}(0 - u) = -\dot{m}u_2$$

$$F = \dot{m}u_2 = (\rho A_1 u_1)\, u_2.$$

Horizontal jet, plane surface inclined to oncoming fluid

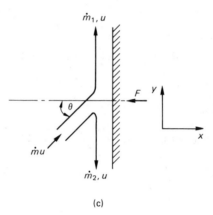

(c)

$$\Sigma F_x = -F = \dot{m}(0 - u\cos\theta)$$

$$F = \dot{m}u\cos\theta = \rho A u^2 \cos\theta.$$

For an ideal fluid the jet speed remains constant after impact.

To find $\dot{m}_1, \dot{m}_2$:

Continuity: $\qquad\qquad\qquad\qquad\qquad \dot{m} = \dot{m}_1 + \dot{m}_2.$

Momentum in y-direction:

$$\Sigma F_y = 0 = (\dot{m}_1 u - \dot{m}_2 u) - \dot{m}u\sin\theta \qquad\qquad \ldots \text{(i)}$$

or $\qquad\qquad\qquad \dot{m}\sin\theta = \dot{m}_1 - \dot{m}_2. \qquad\qquad\qquad \ldots \text{(ii)}$

Solving equations (i) and (ii), we get (adding):

$$\left.\begin{array}{l} 2\dot{m}_1 = \dot{m}(1 + \sin\theta); \\ \text{(subtracting):} \\ 2\dot{m}_2 = \dot{m}(1 - \sin\theta). \end{array}\right\} \quad \text{Hence } \dot{m}_1 \text{ and } \dot{m}_2.$$

Horizontal jet of real fluid, impinging upon a curved surface

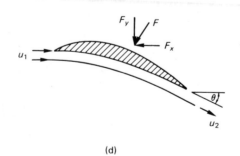

(d)

The jet speed will be reduced as the jet passes over the surface because of viscosity. Let k be the ratio of outlet speed to inlet speed; then

$$u_2 = ku_1.$$

Resolving in x-direction gives

$$F_x = \dot{m}(u_1 - u_2 \cos \theta) = \dot{m}u_1(1 - k \cos \theta).$$

Resolving in y-direction gives

$$F_y = \dot{m}u_2 \sin \theta = \dot{m}u_1 k \sin \theta.$$

Hence

$$F = \sqrt{F_x^2 + F_y^2}$$

$$\alpha = \arctan \frac{F_y}{F_x}.$$

Force Exerted on a Nozzle

To determine the reaction between the fluid and the nozzle and the net force required to hold the nozzle in position when fluid of specific gravity S and at a pressure p just upstream of the nozzle is discharged into the atmosphere: let F be the reaction between the fluid and the nozzle. Then:

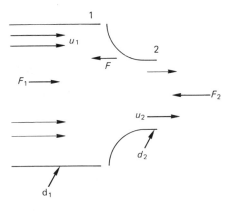

Momentum: $\Sigma F_x = F_1 - F_2 = p_1 A_1 - p_2 A_2 - F = \dot{m}(u_2 - u_1).$

Bernoulli: $\dfrac{u_2^2 - u_1^2}{2g} = \dfrac{p_1 - p_2}{\rho g}.$

Continuity: $A_1 u_1 = A_2 u_2.$

For given values of d_1 and d_2, A_1 and A_2 can be found and u_1 and u_2 can be found from these two since p_1, p_2 and ρ are known. Then $\dot{m} = \rho A_1 u_1$.

Thus F can be calculated from the momentum equation.

Energy Degradation at a Sudden Enlargement in a Pipe

At the enlargement the fluid stream will expand gradually to occupy the new cross-sectional area. In the process, eddies will be formed causing considerable turbulence and energy degradation.

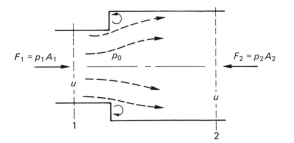

Continuity (ρ constant): $\dot{m} = \rho A_1 u_1 = \rho A_2 u_2.$

Momentum: $p_1 A_1 - p_2 A_2 + p_0 (A_2 - A_1) = \dot{m}(u_2 - u_1).$

However, p_0 is found, experimentally, to equal p_1 approximately.

Thus $(p_1 - p_2) A_2 = \dot{m}(u_2 - u_1) = \rho A_2 u_2 (u_2 - u_1),$

and, rearranging, $(p_2 - p_1) = \rho u_2 (u_1 - u_2)$

or $\dfrac{p_2 - p_1}{\rho g} = \dfrac{u_2 (u_1 - u_2)}{g}.$

Bernoulli (to find the energy degradation at the sudden enlargement):

$$\frac{p_1}{\rho g} + \frac{u_1^2}{2g} = \frac{p_2}{\rho g} + \frac{u_2^2}{2g} + \text{energy degradation};$$

$$\text{energy degradation} = \frac{u_1^2 - u_2^2}{2g} - \left(\frac{p_1 - p_2}{\rho g}\right)$$

and, substituting from above,

$$\text{energy degradation} = \frac{u_1^2 - u_2^2}{2g} - \frac{u_2 (u_1 - u_2)}{g} = \frac{(u_1 - u_2)^2}{2g}.$$

If the exit is to a large container (e.g. a reservoir), $A_2 \to \infty$, $u_2 \to 0$,

$$\text{energy degradation} = \frac{u_1^2}{2g}.$$

Energy Degradation at a Sudden Contraction in a Pipe

Following a sudden contraction, the fluid forms a vena contracta — similar to the flow through an orifice. Energy degradation occurs beyond the vena contracta while the fluid stream expands to occupy the pipe cross-sectional area.

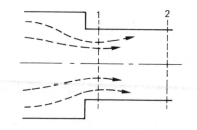

Between sections 1 and 2: $\dfrac{p_2 - p_1}{\rho g} = \dfrac{u_2 (u_1 - u_2)}{g}$

as for a sudden expansion,

$$\text{energy degradation} = \frac{(u_1 - u_2)^2}{2g}.$$

But $A_1 = C_C A_2$ (where C_C is coefficient of contraction)

and since $A_1 u_1 = A_2 u_2$, then $u_1 = \dfrac{u_2}{C_C}.$

Substituting for u_1:
$$\frac{p_2 - p_1}{\rho g} = \frac{u_2^2}{g}\left(\frac{1}{C_C} - 1\right),$$

and
$$\text{energy degradation} = \frac{u_2^2}{2g}\left(\frac{1}{C_C} - 1\right)^2.$$

When $C_C = 0.62$ (as for an orifice),

$$\text{energy degradation} = 0.375\ \frac{u_2^2}{2g}.$$

In practice, however, it has been found that the energy degradation can be as high as $0.5u_2^2/2g$ and this is the value normally assumed.

Mouthpiece

The uniform external mouthpiece

The discharge through an orifice can be increased by fitting a short length of pipe around it. A liquid stream passing through the orifice will contract to the vena contracta and then diverge to occupy the full cross-section of the pipe.

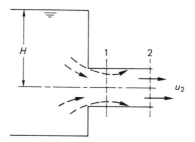

For the enlargement between the vena contracta (1) and the pipe outlet (2),

$$\frac{p_2 - p_1}{\rho g} = \frac{u_2(u_1 - u_2)}{g} = \frac{u_2^2}{g}\left(\frac{1}{C_C} - 1\right).$$

Since p_2 is atmospheric, the pressure head $p_1/\rho g$ will be below atmospheric. Thus the hydrostatic head $(p/\rho g + z)$ between the free surface in the tank and the vena contracta has been increased as a result of the mouthpiece. Applying Bernoulli's equation between the free surface and the pipe outlet and neglecting energy degradation upstream of the vena contracta ($C_u = 1$),

$$H = \frac{u_2^2}{2g} + h_f,$$

where h_f is the energy degradation due to the enlargement between 1 and 2.

$$H = \frac{u_2^2}{2g} + \frac{u_2^2}{2g}\left(\frac{1}{C_C} - 1\right)^2 \qquad \text{and putting} \qquad C_C = 0.62$$

$$= \frac{u_2^2}{2g}(1 + 0.375) = 1.375\ \frac{u_2^2}{2g};$$

i.e. $\quad u_2 = 0.853\sqrt{2gH} \qquad$ (compared with $0.62\sqrt{2gh}$ for orifice).

The convergent–divergent external mouthpiece

This reduces the energy degradation between the vena contracta and the outlet (cf. venturi).

Allowing for friction effects, $u_2 = 0.97\sqrt{2gH}$.

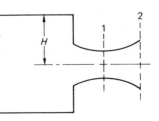

(*Note:* If as a result of attaching a mouthpiece the pressure at the vena contracta is reduced to approximately 0.25 bar, gases dissolved in the fluid may come out of solution and reduce the effectiveness of the mouthpiece.)

5.3 Worked Examples

Example 5.1

Water is discharging vertically downwards from a pipe of diameter 0.2 m into a tank resting on a weighing machine. The pipe outlet is level with the top of the tank which is 3 m high and 1.25 m in diameter. Pitot tube measurements indicate the mean velocity of water at the pipe outlet to be 2.7 m/s.

Determine the percentage error in weighing due to the dynamic force of the jet, if scale readings are taken when the water is initially 0.3 m deep and 2.8 m deep finally.

Answer

Weight of water initially $= m_1 g = \rho V_1 g = \rho\,\dfrac{\pi}{4}\,d^2 L_1 g$

$$= 10^3\,\frac{\text{kg}}{\text{m}^3} \times \frac{\pi}{4} \times 1.25^2\ \text{m}^2 \times 0.3\ \text{m} \times 9.81\,\frac{\text{m}}{\text{s}^2}\left[\frac{\text{N s}^2}{\text{kg m}}\right]$$

$$= 3611.6\ \text{N}.$$

Weight of water finally $= 10^3 \times \frac{1}{4}\pi \times 1.25^2 \times 2.8 \times 9.81 = 33\,708.3\ \text{N}.$

From Bernoulli, $\dfrac{u_{\text{final}}^2}{2g} = \dfrac{u_{\text{initial}}^2}{2g} + h,$

$u_{\text{initial}} = 2.7\,\dfrac{\text{m}}{\text{s}}$ constant.

$h_A = 3\ \text{m} - 0.3\ \text{m} = 2.7\ \text{m},$

$h_B = 3\ \text{m} - 2.8\ \text{m} = 0.2\ \text{m}.$

$$u_{\text{final, A}} = \sqrt{2.7^2\ \frac{m^2}{s^2} + \left(2 \times 9.81\ \frac{m}{s^2} \times 2.7\ m\right)} = 7.763\ \frac{m}{s}\ ,$$

$$u_{\text{final, B}} = \sqrt{2.7^2 + (2 \times 9.81 \times 0.2)} = 3.349\ \frac{m}{s}\ .$$

Force $F_A = -\rho \dot{V}(u_2 - u_1) = 10^3\ \frac{kg}{m}\ (\tfrac{1}{4}\pi \times 0.2^2\ m^2 \times 2.7\ m)\ (7.763 - 0)\ \frac{m}{s}\ \left[\frac{N\ s^2}{kg\ m}\right]$

$$= 658.5\ N,$$

and $\quad F_B = 10^3\ (\tfrac{1}{4}\pi \times 0.2^2 \times 2.7)\ 3.349 = 284.1\ N.$

Thus balance reading (A) = 3611.6 + 658.5 = 4270.1 N
and balance reading (B) = 33 708.3 + 284.1 = 33 992.4 N.
Thus recorded weight = 33 992.4 − 4270.1 = 29 722.3 N.
True weight = 33 708.3 − 3611.6 = 30 096.7 N.

$$\text{Percentage error} = \left(\frac{29\,722.3 - 30\,096.7}{30\,096.7}\right) \times 100 \text{ per cent} = -1.24 \text{ per cent.}$$

Example 5.2

A jet of water flowing in a horizontal plane is divided equally as shown in the diagram by the stationary deflector. Calculate the magnitude and direction of the forces on the deflector in the x- and y-directions. Neglect all energy degradation.

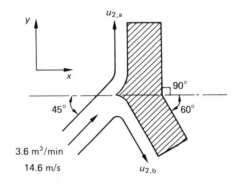

Answer

For no energy degradation, $u_{2,a} = u_{2,b} = u_1 = 14.6\ \frac{m}{s}\ .$

Volume flow rates are: $\dot{V}_{2,a} = \dot{V}_{2,b} = \dfrac{\dot{V}_1}{2} = \dfrac{3.6}{2} = 1.8\ \dfrac{m^3}{s}\ .$

$\Sigma F_x = \rho\ \dfrac{\dot{V}_1}{2}\ (u_{2,b} \cos 60° - u_1 \cos 45°) + \rho\ \dfrac{\dot{V}_1}{2}\ (u_{2,a} \cos 90° - u_1 \cos 45°)$

$\qquad = 10^3\ \dfrac{kg}{m^3} \times 1.8\ \dfrac{m^3}{min}\ \left[\dfrac{min}{60\ s}\right]\ \{(14.6 \cos 60° - 14.6 \cos 45°)$

$\qquad\quad + (0 - 14.6 \cos 45°)\ \dfrac{m}{s}\}\ \left[\dfrac{N\ s^2}{kg\ m}\right]$

$\qquad = -400.4\ N, \qquad$ i.e. from right to left on water.

$$\Sigma F_y = \rho \, \frac{\dot{V}_1}{2} \, (u_{2,a} - u_1 \sin 45°) + \rho \, \frac{\dot{V}_1}{2} \, (u_{2,b} \sin 60° - u_1 \sin 45°)$$

$$= 10^3 \, \frac{\text{kg}}{\text{m}^3} \times 1.8 \, \frac{\text{m}^3}{\text{min}} \left[\frac{\text{min}}{60 \text{ s}} \right] \left\{ (14.6 - 14.6 \sin 45°) \right.$$

$$\left. + (-14.6 \sin 60° - 14.6 \sin 45°) \, \frac{\text{m}}{\text{s}} \right\} \left[\frac{\text{N s}^2}{\text{kg m}} \right]$$

(since vertical component of $u_{2,b}$ is *downwards* and thus negative)

$$= 10^3 \times \frac{1.8}{60} \times 14.6 \left[1 - \frac{1}{\sqrt{2}} - \frac{\sqrt{3}}{2} - \frac{1}{\sqrt{2}} \right] \text{N}$$

$$= -560.7 \text{ N}, \qquad \text{i.e. downwards on water.}$$

Example 5.3

A two-dimensional jet of water impinges on a plane wall at an angle of 60°. The jet is 80 mm thick and its velocity is 10 m/s. Calculate the normal force exerted on the wall per unit width and the thickness of each of the two sheets of water flowing over the wall. Refer to the diagram and neglect the effects of gravity and viscosity.

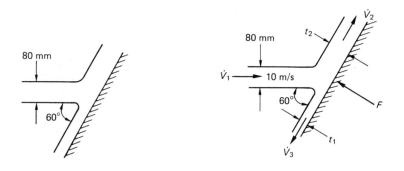

Answer

Force on the wall per unit width:

$$F = \rho \dot{V}_2 u_2 \cos 90° + \rho \dot{V}_3 u_3 \cos 90° - \rho \dot{V}_1 u_1 \sin 60° = -\rho \dot{V}_1 u_1 \sin 60°$$

$$= -10^3 \, \frac{\text{kg}}{\text{m}^3} \left[0.08 \text{ m} \times 1 \text{ m} \times 10 \, \frac{\text{m}}{\text{s}} \right] \left(10 \, \frac{\text{m}}{\text{s}} \times \frac{\sqrt{3}}{2} \right) \left[\frac{\text{N s}^2}{\text{kg m}} \right]$$

$$= -6928 \text{ N}$$

i.e. in a direction diametrically opposite to that in the diagram.
    There is no force parallel to the wall because of shear stresses, and thus no change in momentum in that direction. There is no change in stream velocity if viscosity is negligible.

$$\dot{V}_1 = \dot{V}_2 + \dot{V}_3 \qquad \dots (1); \qquad \text{and hence} \qquad A_1 = A_2 + A_3 \, .$$

Momentum: $\qquad \rho \dot{V}_1 u_1 \cos 60° = \rho \dot{V}_2 u_2 - \rho \dot{V}_3 u_3 \qquad$ (for no change) i.e.

$$\frac{\dot{V}_1}{2} = \dot{V}_2 - \dot{V}_3 \qquad \ldots \text{(ii)}; \qquad \text{and, from (i) and (ii),}$$

$$\dot{V}_2 + \dot{V}_3 = 2(\dot{V}_2 - \dot{V}_3)$$

or
$$\dot{V}_2 = 3\dot{V}_3.$$

Thus
$$t_3 = 20 \text{ mm} \qquad \text{and} \qquad t_2 = 60 \text{ mm.}$$

Example 5.4

A pipe of diameter 0.15 m bends through 90° in a horizontal plane and while bending changes its diameter smoothly to 0.07 m. The pressure in the large end of the pipe is 206 kN/m² gauge. Calculate the magnitude and direction of the resultant horizontal force on the bend when a flow of 0.06 m³/s of water takes place through the pipe. Both inlet and outlet pipes are in the same horizontal plane. Energy degradation may be neglected.

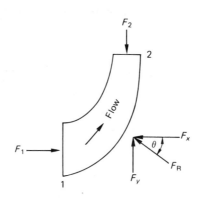

Answer

Bernoulli from 1 to 2:

$$\frac{p_1}{\rho g} + \frac{u_1^2}{2g} + z_1 = \frac{p_2}{\rho g} + \frac{u_2^2}{2g} + z_2 \qquad (z \text{ cancels}).$$

Continuity:
$$u_1 = \frac{\dot{V}}{A_1} = \frac{0.06 \text{ m}^3/\text{s}}{\frac{1}{4}\pi(0.15)^2 \text{ m}^2} = 3.395 \ \frac{\text{m}}{\text{s}} \ ;$$

$$u_2 = \frac{\dot{V}}{A_2} = \frac{0.06}{\frac{1}{4}\pi \times 0.07^2} = 15.59 \ \frac{\text{m}}{\text{s}} \ ,$$

Thus
$$\frac{p_2}{\rho g} = \frac{206\,000 \ \frac{\text{kN}}{\text{m}^2}}{1000 \ \frac{\text{kg}}{\text{m}^3} \times 9.81 \ \frac{\text{m}}{\text{s}^2}} \left[\frac{\text{kg m}}{\text{N s}^2}\right] + \frac{3.395^2 \left(\frac{\text{m}}{\text{s}}\right)^2}{2 \times 9.81 \ \frac{\text{m}}{\text{s}^2}} - \frac{15.59^2 \left(\frac{\text{m}}{\text{s}}\right)^2}{2 \times 9.81 \ \frac{\text{m}}{\text{s}^2}}$$

$$= 21 \text{ m} + 0.587 \text{ m} - 12.388 \text{ m} = 9.2 \text{ m,}$$

$$p_2 = 9.2 \text{ m} \times 10^3 \ \frac{\text{kg}}{\text{m}^3} \times 9.81 \ \frac{\text{m}}{\text{s}^2} \left[\frac{\text{N s}^2}{\text{kg m}}\right] \left[\frac{\text{kN}}{10^3 \text{ N}}\right] = 90.25 \ \frac{\text{kN}}{\text{m}^2} \ .$$

$$F_1 = p_1 A_1 = 206\,000 \ \frac{\text{N}}{\text{m}^2} \times \tfrac{1}{4}\pi \times 0.15^2 \text{ m}^2 = 3640 \text{ N,}$$

$$F_2 = p_2 A_2 = 90\,250 \times \tfrac{1}{4}\pi \times 0.07^2 = 347.3 \text{ N.}$$

Momentum: $-F_x + F_1 = \rho \dot{V}(u_{2x} - u_{1x})$,

$$-F_x + 3640 \text{ N} = 10^3 \frac{\text{kg}}{\text{m}^3} \times 0.06 \frac{\text{m}^3}{\text{s}} (0 - 3.395) \frac{\text{m}}{\text{s}} \left[\frac{\text{N s}^2}{\text{kg m}} \right],$$

$$-F_x = -3640 - 203.7 = -3843.7 \text{ N},$$

$$F_x = 3843.7 \text{ N right to left on the water.}$$

$$F_y - F_2 = \rho \dot{V}(u_{2y} - u_{1y}),$$

$$F_y = 347.3 \text{ N} + 10^3 \frac{\text{kg}}{\text{m}^3} \times 0.06 \frac{\text{m}^3}{\text{s}} (15.59 - 0) \frac{\text{m}}{\text{s}} \left[\frac{\text{N s}^2}{\text{kg m}} \right]$$

$$= 1282.7 \text{ N upwards in water.}$$

Thus $\quad F_R = \sqrt{F_y^2 + F_x^2} = \sqrt{3843.7^2 + 1282.7^2} \text{ N} = 4052 \text{ N}.$

$$\theta = \arctan \frac{F_y}{F_x} = \arctan \frac{1282.7}{3843.7} = 18.45°.$$

Thus reaction force on bend $= -F_R$ at θ to horizontal as shown.

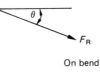

On bend

Example 5.5

A conical enlargement in a vertical pipeline is 1.5 m long and enlarges the pipe diameter from 0.3 m at the lower end to 0.6 m at the larger end. Calculate the magnitude and direction of the vertical force on the enlargement when a flow of 0.28 m³/s of water passes upward through the pipeline and the pressure at the smaller end of the enlargement is 207 kN/m² gauge. Assume that the energy degradation in the enlargement may be expressed as

$$\frac{0.2}{2g} (u_1 - u_2)^2$$

where u_1 and u_2 are the flow velocities at inlet and outlet respectively.

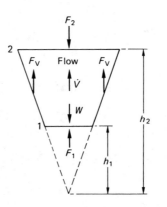

Answer

Bernoulli from 1 to 2: $\dfrac{p_1}{\rho g} + \dfrac{u_1^2}{2g} + z_1 = \dfrac{p_2}{\rho g} + \dfrac{u_2^2}{2g} + z_2 + h_f$ $\quad (z_2 - z_1 = 1.5 \text{ m}).$

Continuity: $\quad u_1 = \dfrac{\dot{V}}{A_1} = \dfrac{0.28 \,\dfrac{\text{m}^3}{\text{s}}}{\frac{1}{4}\pi\,(0.3^2)\,\text{m}^2} = 3.96\,\dfrac{\text{m}}{\text{s}}\,.$

$\left. \begin{array}{c} \\ \\ \end{array} \right\}$

$u_2 = \dfrac{0.28}{\frac{1}{4}\pi\,(0.6^2)} = 0.99\,\dfrac{\text{m}}{\text{s}}\,.$ $\qquad (u_1 - u_2) = 2.97\,\dfrac{\text{m}}{\text{s}}\,.$

$$\dfrac{p_2}{\rho g} = \dfrac{207\,000\,\dfrac{\text{kN}}{\text{m}^2}}{10^3\,\dfrac{\text{kg}}{\text{m}^3} \times 9.81\,\dfrac{\text{m}}{\text{s}^2}}\left[\dfrac{\text{kg m}}{\text{N s}^2}\right] + \dfrac{3.96^2\,\left(\dfrac{\text{m}}{\text{s}}\right)^2}{2 \times 9.81\,\dfrac{\text{m}}{\text{s}^2}} - 1.5\,\text{m} - \dfrac{0.99^2\,\left(\dfrac{\text{m}}{\text{s}}\right)^2}{2 \times 9.81\,\dfrac{\text{m}}{\text{s}^2}}$$

$$- \dfrac{0.2 \times 2.97^2\,\left(\dfrac{\text{m}}{\text{s}}\right)^2}{2 \times 9.81\,\dfrac{\text{m}}{\text{s}^2}}$$

$\qquad = 21.1\,\text{m} + 0.8\,\text{m} - 1.5\,\text{m} - 0.05\,\text{m} - 0.09\,\text{m} = 20.26\,\text{m},$

$p_2 = 20.26\,\text{m} \times 10^3\,\dfrac{\text{kg}}{\text{m}^3} \times 9.81\,\dfrac{\text{m}}{\text{s}^2}\left[\dfrac{\text{N s}^2}{\text{kg m}}\right] = 198\,750\,\dfrac{\text{N}}{\text{m}^2}\,.$

$F_1 = p_1 A_1 = 207\,000\,\dfrac{\text{N}}{\text{m}^2} \times \frac{1}{4}\pi \times 0.3^2\,\text{m}^2 = 14\,632\,\text{N},$

$F_2 = p_2 A_2 = 198\,750 \times \frac{1}{4}\pi \times 0.6^2 = 56\,195\,\text{N}.$

Momentum: $\qquad \Sigma F_V = F_1 + F_V - F_2 - W = \rho \dot{V}(u_2 - u_1),$

$\qquad\qquad\qquad W = \rho g \frac{1}{3}(\frac{1}{4}\pi)[D_2^2 h_2 - D_1^2 h_1].$

Now $\qquad\qquad\qquad \dfrac{h_1}{h_1 + 1.5} = \frac{1}{2} \qquad$ (similar Δs);

$\qquad\qquad\qquad\qquad h_1 = 1.5\,\text{m}; h_2 = 3\,\text{m};$

$W = \frac{1}{12}\pi \times 10^3\,\dfrac{\text{kg}}{\text{m}^3} \times 9.81\,\dfrac{\text{m}}{\text{s}^2}\,[(0.6^2 \times 3) - (0.3^2 \times 1.5)]\,\text{m}^3\left[\dfrac{\text{N s}^2}{\text{kg m}}\right]$

$\qquad = 2427\,\text{N}.$

$F_V = 10^3\,\dfrac{\text{kg}}{\text{m}^3} \times 0.28\,\dfrac{\text{m}^3}{\text{s}}\,(-2.97)\,\dfrac{\text{m}}{\text{s}}\left[\dfrac{\text{N s}^2}{\text{kg m}}\right] + 2427\,\text{N} + 56\,195\,\text{N} - 14\,632\,\text{N}$

$\qquad = -43\,158\,\text{N} \qquad$ (upwards on water).

Example 5.6

A nozzle of diameter 0.1 m is bolted with 8 bolts to the flange of a horizontal pipeline of diameter 0.3 m and discharges water into the atmosphere. Calculate the load in each bolt when the pressure in the pipe is 550 kN/m² gauge. Assume that for the nozzle, $C_u = 0.98$, $C_C = 1.0$.

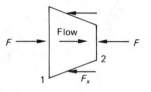

Answer

Bernoulli from 1 to 2:

$$\frac{p_1}{\rho g} + \frac{u_1^2}{2g} + z_1 = \frac{p_2}{\rho g} + \frac{u_2^2}{2g} + z_2 \qquad (z_2 - z_1 = 0).$$

Continuity: $A_2 u_2 = A_1 u_1$;

$$u_1 = \frac{u_2}{9} \qquad \left(\text{since } D_2 = \frac{D_1}{3}\right);$$

$$\text{energy degradation} = \frac{u_2^2}{2g}\left(\frac{1}{C_u^2} - 1\right) = 0.04\,\frac{u_2^2}{2g}.$$

Thus $\dfrac{550\,000\ \dfrac{N}{m^2}}{10^3\ \dfrac{kg}{m^3} \times 9.81\ \dfrac{m}{s^2}}\left[\dfrac{kg\ m}{N\ s^2}\right] + \dfrac{u_1^2}{2 \times 9.81\ \dfrac{m}{s^2}} = 9^2\ \dfrac{u_1^2}{2 \times 9.81\ \dfrac{m}{s^2}}\,[1 + 0.04],$

giving $\qquad u_1 = 3.64\ \dfrac{m}{s}; \qquad u_2 = 32.72\ \dfrac{m}{s};$

and $\qquad \dot{V} = A_1 u_1 = \dfrac{1}{4}\pi \times 0.3^2\ m^2 \times 3.64\ \dfrac{m}{s} = 0.257\ \dfrac{m^3}{s}.$

$$F_1 = p_1 A_1 = 550\,000\ \frac{N}{m^2} \times \frac{1}{4}\pi(0.3^2)\ m^2 = 38\,877\ N;$$

$F_2 = 0$ (atmospheric pressure).

Momentum: $\Sigma F_x = F_1 - F_2 - F_x = \rho\dot{V}(u_2 - u_1)$;

$$-F_x = 10^3\ \frac{kg}{m^3} \times 0.257\ \frac{m^3}{s}(32.72 - 3.64)\ \frac{m}{s}\left[\frac{N\ s^2}{kg\ m}\right] - 38\,877\ N,$$

$$F_x = 31\,403\ N \text{ for 8 bolts,}$$

$$\frac{F_x}{8} = 3925.4\ N \text{ per bolt.}$$

Example 5.7

Water flows at the rate of 0.075 m³/s along a pipeline in which there is a sudden change of section from a diameter of 0.25 m to a diameter of 0.35 m. Determine the energy degradation, the pressure head difference across the change of section and the power required to overcome the energy degradation if the flow is from the smaller to the larger diameter.

Recalculate the power if the flow is in the other direction. Take $C_C = 0.6$.

Vena contracta

0.35 m dia. → 0.25 m dia.

Answer

Flow from 2 to 1:
$$h_f = \frac{(u_2 - u_1)^2}{2g}.$$

Continuity:
$$u_2 = \frac{\dot{V}}{A_2} = \frac{0.075\ \frac{m^3}{s}}{\frac{1}{4}\pi \times 0.25^2\ m^2} = 1.528\ \frac{m}{s}\ ;$$

$$u_1 = \frac{\dot{V}}{A_1} = \frac{0.075}{\frac{1}{4}\pi \times 0.35^2} = 0.78\ \frac{m}{s}.$$

$$h_f = \frac{(1.528 - 0.78)^2}{2 \times 9.81\ \frac{m}{s^2}}\ \frac{m^2}{s^2} = 0.0286\ m = 28.6\ mm\ water.$$

$$\frac{\Delta p_{21}}{\rho g} = \frac{u_1(u_2 - u_1)}{g} = \frac{0.78\ \frac{m}{s}\ (1.528 - 0.78)\ \frac{m}{s}}{9.81\ \frac{m}{s^2}} = 0.0595\ m\ (59.5\ mm\ water).$$

$$\dot{W} = \dot{V} \times \rho g h_f = 0.075\ \frac{m^3}{s} \times 10^3\ \frac{kg}{m^3} \times 9.81\ \frac{m}{s^2} \times 0.0286\ m\ \left[\frac{N\ s^2}{kg\ m}\right]\left[\frac{W\ s}{N\ m}\right]$$

$$= 21.04\ W.$$

Flow from 1 to 2:
$$u_3 = \frac{\dot{V}}{A_3} = \frac{\dot{V}}{C_C A_2} = \frac{u_2}{0.6} = 2.55\ \frac{m}{s}\ ;$$

$$h_f = \frac{(u_3 - u_2)^2}{2g} = \frac{(2.55 - 1.528)\ \frac{m^2}{s^2}}{2 \times 9.81\ \frac{m}{s^2}} = 0.053\ m = 53\ mm\ water.$$

$$\dot{W} = \dot{V}\rho g h = 0.075 \times 10^3 \times 9.81 \times 0.053 = 38.99\ W\ (with\ usual\ units).$$

Example 5.8

A pipe of diameter 0.1 m is joined in line with one of diameter 0.15 m. Between the connecting flanges a plate is bolted having a sharp-edged orifice of diameter 60 mm. Water flows at the rate of 0.34 m^3/min from the smaller section to the larger. Assuming that the value of C_u is unity estimate the pressure difference between the two pipes. Take C_C for the orifice as 0.61 and assume that the velocity is uniform across the pipes where tappings are made.

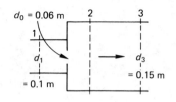

Answer

Continuity: $u_1 = \dfrac{\dot{V}}{A_1} = \dfrac{0.34 \dfrac{m^3}{mm} \left[\dfrac{min}{60\ s}\right]}{\frac{1}{4}\pi \times 0.1^2\ m^2} = 0.722\ \dfrac{m}{s}$;

$u_2 = \dfrac{\dot{V}}{A_2} = \dfrac{\dot{V}}{C_c A_0} = \dfrac{0.34}{60 \times 0.61 \times \frac{1}{4}\pi \times 0.06^2} = 3.285\ \dfrac{m}{s}$;

$u_3 = \dfrac{\dot{V}}{A_3} = \dfrac{0.34}{60 \times \frac{1}{4}\pi \times 0.15^2} = 0.321\ \dfrac{m}{s}$.

$h_{f,23} = \dfrac{(u_2 - u_3)^2}{2g} = \dfrac{(3.285 - 0.321)^2\ \dfrac{m^2}{s^2}}{2 \times 9.81\ \dfrac{m}{s^2}} = 0.448\ m.$

Bernoulli from 1 to 3: $\dfrac{p_1}{\rho g} + \dfrac{u_1^2}{2g} = \dfrac{p_3}{\rho g} + \dfrac{u_2^2}{2g} + \dfrac{(u_2 - u_3)^2}{2g}$.

Thus $\dfrac{p_1 - p_3}{\rho g} = \dfrac{(u_2 - u_3)^2}{2g} + \dfrac{u_3^2}{2g} - \dfrac{u_1^2}{2g}$

$= 0.448\ m + \dfrac{0.321^2\ \dfrac{m^2}{s^2}}{2 \times 9.81\ \dfrac{m}{s^2}} - \dfrac{0.722^2\ \dfrac{m^2}{s^2}}{2 \times 9.81\ \dfrac{m}{s^2}}$

$= 0.448 + 0.0053 - 0.0256$

$= 0.4277\ m\ water;$

and $p_1 - p_3 = 10^3\ \dfrac{kg}{m^3} \times 9.81\ \dfrac{m}{s^2} \times 0.4277\ m \left[\dfrac{N\ s^2}{kg\ m}\right] \left[\dfrac{kN}{10^3\ N}\right]$

$= 4.196\ \dfrac{kN}{m^2}$.

Example 5.9

Water discharges from a tank through an orifice of diameter 75 mm in its side. Determine the rate of discharge when the head over the orifice is 1.25 m and $C_D = 0.6$.

An external mouthpiece in the form of a length of pipe of inside diameter 75 mm is fitted around the orifice. Within the mouthpiece the water flow contracts to 0.6 times the cross-sectional area of the pipe, after which it re-expands to fill the tube. Determine the percentage increase in flow obtained by introducing the mouthpiece.

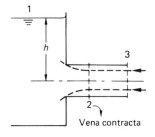

Vena contracta

Answer

Without mouthpiece:

$$\dot{V} = A_o C_D \sqrt{2gh} = \tfrac{1}{4} \pi \times 0.075^2 \text{m}^2 \times 0.6 \sqrt{2 \times 9.81 \times 1.25} \ \frac{\text{m}}{\text{s}}$$

$$= 0.0131 \ \frac{\text{m}^3}{\text{s}} \ .$$

With mouthpiece:

$$h_{f,23} = \frac{(u_2 - u_3)^2}{g} = \frac{u_2^2}{2g}\left(\frac{1}{0.6} - 1\right)^2 = \frac{4}{9}\frac{u_3^2}{2g} .$$

Bernoulli from 1 to 3:

$$\frac{p_1}{\rho g} + \frac{u_1^2}{2g} + z_1 = \frac{p_3}{\rho g} + \frac{u_3^2}{2g} + z_3 + h_f$$

and

$$z_1 - z_3 = h.$$

Since 1 is a free surface,

$$h = \frac{u_3^2}{2g} + \frac{4}{9}\frac{u_3^2}{2g} = \frac{13}{9}\frac{u_3^2}{2g}$$

and

$$\dot{V} = A_3 u_3 = A_o u_3 = A_o \sqrt{2gh\,(9/13)}.$$

Thus

$$\frac{\dot{V} \text{ with mouthpiece}}{\dot{V} \text{ without mouthpiece}} = \frac{A_o \sqrt{2gh}}{A_o \sqrt{2gh}} \frac{\sqrt{9/13}}{C_D} = \frac{\sqrt{9/13}}{0.6}$$

$$\frac{\dot{V} \text{ with}}{\dot{V} \text{ without}} = 1.387,$$

i.e. there is 38.7 per cent increase in flow.

Example 5.10

A sharp-edged orifice of area A is situated in the bottom of a vessel containing water to a depth h. The orifice is equipped with an external tube of the same area A and length L. After passing through the plane of the orifice the water contracts to a diameter equal to 0.6 times A, after which it re-expands to fill the external tube. Estimate the rate of discharge in terms of h and L, given that viscous friction at the orifice is negligible. Assuming that the vena contracta is in the plane of the orifice, determine the pressure there relative to the atmosphere.

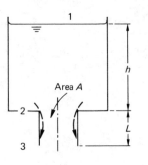

Answer

$$h_{f,23} = \frac{(u_2 - u_3)^2}{2g} = \frac{u_3^2}{2g}\left(\frac{1}{0.6} - 1\right)^2 = \frac{4}{9}\frac{u_3^2}{2g}.$$

Bernoulli from 1 to 3: $\dfrac{p_1}{\rho g} + \dfrac{u_1^2}{2g} + z_1 = \dfrac{p_3}{\rho g} + \dfrac{u_3^2}{2g} + z_3 + h_{f,23},$

$$z_1 - z_3 = h + L.$$

Thus, since 1 is a free surface,

$$h + L = \frac{u_3^2}{2g}\left(1 + \frac{4}{9}\right) = \frac{13}{9}\frac{u_3^2}{2g};$$

$$u_3 = \sqrt{\frac{9}{13}\,2g\,(h + L)};$$

$$\dot{V} = A_3 u_3 = A\sqrt{\frac{9}{13}\,2g\,(h + L)} = 3\sqrt{\frac{2g\,(h + L)}{13}}.$$

Bernoulli from 1 to 2: $\dfrac{p_1}{\rho g} + \dfrac{u_1^2}{2g} + z_1 = \dfrac{p_2}{\rho g} + \dfrac{u_2^2}{2g} + z_2;$

and, since $u_1 = 0$,

$$\frac{p_1 - p_2}{\rho g} = -(z_1 - z_2) + \frac{u_2^2}{2g} - \frac{u_1^2}{2g} = -h + \frac{u_2^2}{2g} = -h + \frac{u_3^2}{2g}\times\frac{1}{0.6^2}$$

$$= -h + \frac{1}{0.36}\times\frac{9}{13}\,(h + L) = -h + \frac{25}{13}\,(h + L)$$

$$= \frac{12h + 25L}{13}.$$

Example 5.11

A convergent–divergent nozzle is fitted into the side of an open tank containing water and, under a constant head of 2.5 m above the centre-line of the nozzle, discharges to the atmosphere.

Calculate the ratio of the exit diameter to the throat diameter of the nozzle for maximum discharge, making the following assumptions:
(a) the height of the water barometer is 10.4 m,
(b) separation of dissolved gases will occur at an absolute head of 2.2 m,
(c) the only hydraulic energy degradation occurs in the divergent portion of the nozzle and amounts to 20 per cent of the energy degradation at a sudden enlargement for the same change in area.

If the throat diameter is 40 mm calculate the maximum discharge.

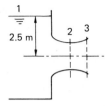

Answer

$$\frac{p_1}{\rho g} = 10.4 \text{ m water}; \qquad \left(\frac{p_2}{\rho g}\right)_{\text{m in}} = 2.2 \text{ m water}.$$

Bernoulli from 1 to 2:

$$\frac{p_1}{\rho g} + \frac{u_1^2}{2g} + z_1 = \frac{p_2}{\rho g} + \frac{u_2^2}{2g} + z_2 \qquad (u_1 = 0),$$

or

$$10.4 \text{ m } + 2.5 \text{ m} = 2.2 \text{ m} + \frac{u_2^2}{2g}.$$

Thus

$$\frac{u_2^2}{2g} = 10.7 \text{ m water}.$$

Bernoulli from 1 to 3:

$$\frac{p_1}{\rho g} + \frac{u_2^2}{2g} + z_1 = \frac{p_3}{\rho g} + \frac{u_3^2}{2g} + z_3 + 0.2 \frac{(u_2 - u_3)^2}{2}$$

or

$$10.4 \text{ m} + 2.5 - 10.4 = \frac{u_3^2}{2g} + \frac{0.2 (u_2 - u_3)^2}{2g} = 2.5 \text{ m water}.$$

Thus

$$1.2 \frac{u_3^2}{2g} + 0.2 (10.7) - 0.4 \sqrt{10.7} \left(\frac{u_3^2}{2g}\right)^{1/2} = 2.5,$$

$$1.2 \left(\frac{u_3^2}{2}\right) - 1.308 \left(\frac{u_3^2}{2}\right)^{1/2} - 0.36 = 0;$$

or

$$\left(\frac{u_3^2}{2g}\right)^{1/2} = \frac{1.308 \pm \sqrt{1.308^2 + 4(1.2)(0.36)}}{2 \times 1.2} = 1.318 \text{ m}^{1/2}$$

Thus

$$\frac{u_2}{u_3} = \frac{\sqrt{10.7}}{1.318} = 2.482.$$

Continuity:

$$\frac{d_3}{d_2} = \sqrt{\frac{u_2}{u_3}} = \sqrt{2.482} = 1.575.$$

$$\hat{V} = A_2 \hat{u}_2 = \tfrac{1}{4} \pi \times 0.04^2 \sqrt{10.7 \times 2 \times 9.81} \frac{\text{m}^3}{\text{s}} = 0.0182 \frac{\text{m}^3}{\text{s}}.$$

Exercises

1 A conical reducer in a vertical pipe is 3 m long and reduces the pipe diameter from 0.6 m at the lower end to 0.3 m. Calculate the magnitude and direction of the vertical force on this conical section when a flow of 0.633 m³/s of water

flows upwards through the pipeline and the pressure at inlet to the reducer is 20.6 kN/m² gauge. Neglect all energy degradation through the section.

(*Answer:* 300 N upwards)

2 A nozzle of exit diameter 25 mm discharges water at the rate of 0.014 m³/s in a horizontal plane. The jet of water is deflected through an angle of 120° by a stationary rough curved vane and then strikes a fixed flat plate placed normal to the oncoming jet and in the same horizontal plane (see diagram).

If the force which the water jet exerts on the flat plate is 350 N, determine the magnitude and direction of the resultant force on the curved vane. Assume that C_C = 1.0 for the nozzle.　　　　　　(*Answer:* 650 N, 27.7 to the horizontal)

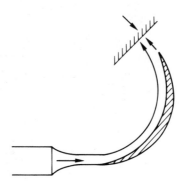

3 By applying the momentum equation to a fluid passing through a sudden enlargement it can be shown that the energy degradation is given by

$$h_f = (u_1 - u_2)^2/2g,$$

where u_1 and u_2 are the upstream and downstream fluid velocities respectively. By applying this result to flow through a sudden contraction, obtain an expression for the energy degradation in terms of C_C, the coefficient of contraction.

A tank of water is emptied through an orifice in its side. In order to increase the flow, a uniform mouthpiece is fitted horizontally around the orifice and the emerging jet of water has the same diameter as the original orifice.

If C_C for the orifice alone is 0.62 and for the vena contracta in the mouthpiece when fitted is 0.5, determine the percentage increase in flow rate.

On the basis of increasing the flow rate from a tank, say, briefly with reasons, how you would rate the mouthpieces shown in the diagram.

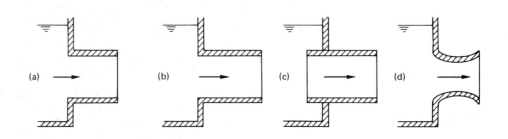

(*Answer:* 14 per cent increase with the vena contracta)

4 Show that when an incompressible fluid passes through a sudden enlargement in a pipe the energy degradation is given by

$$h_f = (u_1 - u_2)^2 / 2g,$$

where u_1 and u_2 are the upstream and downstream velocities respectively.

Water flows at the rate of 0.1 m³/s along a pipe of diameter 0.25 m in which there is a sudden enlargement to 0.4 m diameter. Determine the power required to overcome the energy degradation at this sudden change of section.

(Answer: 77.1 watts)

6 Flow in Pipes of Viscous Incompressible Fluids

6.1 Introduction

The flow of fluids through pipes may be either laminar or turbulent as defined on page xiv. Flow in a closed pipe results from a pressure difference between inlet and outlet. The pressure is affected by fluid properties and flow rate. When the flow velocity is significant compared with the sonic velocity of the fluid medium (as in compressible flows) the sonic velocity must also be included.

By the technique of dimensional analysis (treated in the next chapter) it can be shown that a general pipe flow problem can be summarised as

$$\Delta p = \phi\, (\rho, u, \mu, D, L, u_a),$$

where Δp = pressure loss between pressure tappings,
ρ = density of fluid,
u = velocity of fluid,
μ = viscosity of fluid (absolute),
D = diameter of pipe,
L = pipe length between pressure tappings,
u_a = acoustic velocity.

A dimensional analysis reveals that

$$\frac{\Delta p}{\rho u^2} = \phi\left[\frac{\rho u D}{\mu}, \frac{u}{u_a}, \frac{L}{D}\right],$$

where the term on the left-hand side is called the pressure coefficient (when multiplied by $\frac{1}{2}$ to make the denominator a kinetic-energy term). The term $(\rho u D)/\mu$ is called the Reynolds number of the flow and (u/u_a) is the Mach number of the flow.

It may easily be shown that each of these terms is a ratio of forces. In particular, the Reynolds number, which is of significance in this chapter, may be written as

$$\frac{\rho u D}{\mu} \equiv \frac{\dfrac{M}{L^3}\dfrac{L}{t}L}{\dfrac{M}{Lt}} \equiv \frac{\left[\dfrac{Ft^2}{ML}\right]}{\left[\dfrac{Ft^2}{ML}\right]} \equiv \frac{\dfrac{Ft}{L^2}}{\dfrac{Ft}{L^2}}.$$

The numerator is an inertia force and the denominator is a viscous force. Reynolds number is thus the ratio of inertia forces to viscous forces in the flow. It can be shown that the first term is the ratio of pressure forces to inertia forces and Mach number is the ratio of dynamic force at flow velocity to the dynamic force at sonic velocity.

When a real fluid flows along a pipe it experiences a resistance to flow due to the viscous shear stress between the fluid and the pipe wall.

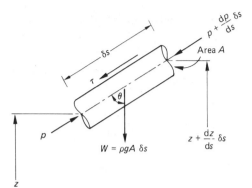

Applying Newton's second law of motion to the element of fluid shown in the diagram, we get

$$\text{force} = \text{rate of change of momentum},$$

and, resolving along the streamline,

$$\left[p - \left(p + \frac{\mathrm{d}p}{\mathrm{d}s} \, \delta s \right) \right] A - \rho g A \, \delta s \cos \theta - \tau P \, \delta s = m \, \frac{\mathrm{d}u}{\mathrm{d}t}$$

where P is the wetted perimeter.

Referring to Chapter 3, page 32, in deriving Euler's equation we use the total derivative

$$\frac{\mathrm{d}u}{\mathrm{d}t} \quad \text{defined as} \quad \frac{\mathrm{d}u}{\mathrm{d}s} \times \frac{\mathrm{d}s}{\mathrm{d}t}, \qquad \text{i.e.} \qquad u \, \frac{\mathrm{d}u}{\mathrm{d}s}.$$

Furthermore, $$\cos \theta = \frac{\mathrm{d}z}{\mathrm{d}s}.$$

Also, $$m = \rho A \, \delta s \qquad \text{(mass of element)}.$$

Thus $$\frac{\mathrm{d}p}{\mathrm{d}s} \, A \, \delta s - \rho g A \, \delta s \, \frac{\mathrm{d}z}{\mathrm{d}s} - \tau P \, \delta s = \rho A \, \delta s \, u \, \frac{\mathrm{d}u}{\mathrm{d}s}.$$

Dividing by $\rho g A \, \delta s$ and rearranging gives

$$\frac{1}{\rho g} \, \frac{\mathrm{d}p}{\mathrm{d}s} + \frac{\mathrm{d}z}{\mathrm{d}s} + \frac{\tau P}{\rho A g} + \frac{u}{g} \, \frac{\mathrm{d}u}{\mathrm{d}s} = 0.$$

For an incompressible fluid ($\rho = $ constant), and assuming that the shear stress τ does not vary with s as in a uniform pipe, integrating with respect to s,

$$\frac{p}{\rho g} + z + \frac{\tau P s}{\rho A g} + \frac{u^2}{2g} = \text{constant}.$$

Thus between points 1 and 2 distance L apart,

$$\frac{p_1}{\rho g} + z_1 + \frac{\tau P s_1}{\rho A g} + \frac{u_1^2}{2g} = \frac{p_2}{\rho g} + z_2 + \frac{\tau P s_2}{\rho A g} + \frac{u_2^2}{2g}.$$

Now $s_1 - s_2 = L$, and thus

$$\frac{p_1}{\rho g} + z_1 + \frac{u_1^2}{2g} = \frac{p_2}{\rho g} + z_2 + \frac{u_2^2}{2g} + \frac{\tau P L}{\rho A g},$$

where $\dfrac{\tau P L}{\rho A g} = h_\mathrm{f}$, the energy degradation due to friction.

Defining the coefficient of friction $f = \dfrac{\tau}{\frac{1}{2}\rho u^2}$ and substituting for τ,

$$h_f = \tfrac{1}{2}\rho u^2\, f\, \frac{PL}{\rho A g} = f\, \frac{PL}{A}\, \frac{u^2}{2g}.$$

For circular pipes, $P = \pi D$,

$$A = \frac{\pi D^2}{4},$$

giving $\qquad\qquad h_f = \dfrac{4fL}{D}\, \dfrac{u^2}{2g} \qquad$ (Darcy formula).

This result holds for both laminar and turbulent flow and we need to select an appropriate value of f from some source of information determined by experiment for each given case (e.g. a Moody diagram).

For non-circular pipes an equivalent diameter D_e is used such that

$$\frac{P}{A} = \frac{4}{D_e}, \qquad \text{i.e.} \qquad D_e = \frac{4A}{P} \qquad \text{and} \qquad h_f = \frac{4fL}{D_e} \times \frac{u^2}{2g}.$$

6.2 Flow Between Two Reservoirs

Energy Degradation

Where a fluid flows between two reservoirs, mechanical energy degradation due to friction will occur chiefly in the pipe but also at sudden changes of section, e.g. at pipe inlet and outlet.

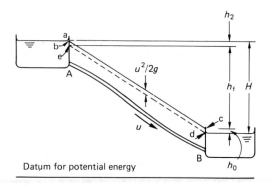

Datum for potential energy

Bernoulli's equation between the two reservoir surfaces in the diagram:

$$\frac{p_1}{\rho g} + \frac{u_1^2}{2g} + z_1 = \frac{p_2}{\rho g} + \frac{u_2^2}{g} + z_2 + \text{energy degradation}.$$

But $\qquad\qquad p_1 = p_2, \qquad u_1 = u_2 = 0,$

$$z_1 - z_2 = H = \text{total degradation} = h_i + h_f + h_o,$$

where $\;h_i$ = inlet degradation = $0.5u^2/2g$ (for sudden contraction),

$\qquad h_f$ = pipe friction degradation = $4\,\dfrac{fL}{D}\,\dfrac{u^2}{2g}$,

$\qquad h_o$ = exit degradation = $u^2/2g$ (sudden expansion),

i.e. $\qquad H = \dfrac{u^2}{2g}\left(1.5 + \dfrac{4fL}{D}\right).$

For long pipes the term $(4fL/D)$ is very large compared with 1.5, and in such cases the inlet and outlet degradation can be neglected.

$$H = \frac{4fL}{D} \frac{u^2}{2g} = h_f.$$

e.g. for a pipe of diameter 1 m, 10 000 m long in which $f = 0.006$,

$$h_f = \frac{4fL}{D} \frac{u^2}{2g} = \frac{4 \times 0.006 \times 10\,000}{1} \times \frac{u^2}{2g} = 240\, \frac{u^2}{2g}.$$

Hydraulic Gradient

In the diagram on page 78, abcd represents the total head and AB the pipeline represents the potential head at any point. The difference between them will represent the sum of the pressure and kinetic heads. If a line ed is drawn below the total head line and distance $u^2/2g$ from it then the difference between ed and AB will represent the pressure head. The gradient ed is called the hydraulic gradient.

For a uniform pipe ed is parallel to bc and the vertical distance between e and d is h_f. If L is the distance between reservoirs then the hydraulic gradient

$$i = \frac{h_f}{L}.$$

In long pipes, $\qquad\qquad h \approx H; \qquad i = \frac{H}{L}.$

Vertical irregularities in pipelines (e.g. hills) are usually small compared with horizontal distances. Pipe length can be taken as the horizontal distance between reservoirs.

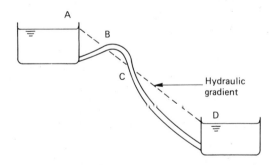

The pressure head in a pipeline is equal to the vertical distance between the pipe centre-line and the hydraulic gradient. Between A and B, C and D the pressure head is positive, i.e. the pressure is above atmospheric. At B and C the pressure is atmospheric. Between B and C the pressure head is negative and the pressure is below atmospheric.

A pipeline which rises above its hydraulic gradient is known as a siphon. At pressures below about 2.5 m water, gases will separate out from the water sufficiently to cause a break in the flow. At no section of the pipe should the pressure be less than 2.5 m of water absolute or 7.5 m water vacuum, i.e. no pipe should lie more than 7.5 m above its hydraulic gradient.

Pressure at a Point in a Pipeline

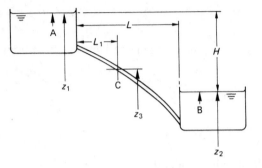

To find the pressure at an intermediate point C:

Apply Bernoulli's equation between reservoirs to find the pipe velocity:

$$H = \left(1.5 + \frac{4fL}{D}\right)\frac{u^2}{2g}.$$

Applying Bernoulli between A and C gives

$$\frac{p_1}{\rho g} + \frac{u_1^2}{2g} + z_1 = \frac{p_2}{\rho g} + \frac{u_3^2}{2g} + z_3 + h_{f,13}$$

and since $u_1 = 0$, $u_3 = u$,

$$\frac{p_3 - p_1}{\rho g} = (z_1 - z_3) - \left(1.5 + \frac{4fL}{D}\ \frac{u^2}{2g}\right) \qquad \text{(hence } p_3\text{)}.$$

Pipes in Series

Applying Bernoulli's equation between A and B gives

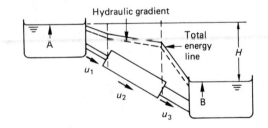

$$\frac{p_1}{\rho g} + \frac{u_1^2}{2g} + z_1 = \frac{p_2}{\rho g} + \frac{u_2^2}{2g} + z_2 + \Sigma \text{ energy degradation,}$$

giving $H = \Sigma$ energy degradation, which includes distributed pipe friction effects and friction due to sudden changes of section.

i.e.
$$H = \frac{0.5u_1^2}{2g} + \frac{(u_1 - u_2)^2}{2g} + \frac{1.5u_3^2}{g} + \sum \frac{4fL}{D}\ \frac{u^2}{2g}.$$

Pipes in Parallel

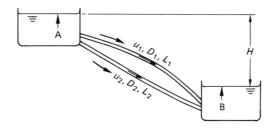

Applying Bernoulli between A and B for each pipe gives

$$H = \left(1.5 + \frac{4fL_1}{D_1}\right)\frac{u_1^2}{2g} = \left(1.5 + \frac{4fL_2}{D_2}\right)\frac{u_2^2}{2g}.$$

For long pipes,
$$H = \frac{4fL_1}{D_1}\frac{u_1^2}{2g} = \frac{4fL_2}{D_2}\frac{u_2^2}{2g}.$$

Pipes in Series and Parallel

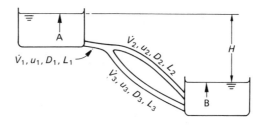

Neglecting all degradation of energy except pipe friction, apply Bernoulli between A and B. The degradation due to friction is the same through pipes 1 and 2 or pipes 1 and 3.

i.e.
$$H = \frac{4fL_1}{D_1}\frac{u_1^2}{2g} + \frac{4fL_2}{D_2}\frac{u_2^2}{2g} = \frac{4fL_1}{D_1}\frac{u_1^2}{2g} + \frac{4fL_3}{D_3}\frac{u_2^2}{2g} \qquad \ldots (1)$$

and hence
$$\frac{4fL_2}{D_2}\frac{u_2^2}{2g} = \frac{4fL_3}{D_3}\frac{u_2^2}{2g}. \qquad \ldots (2)$$

Continuity gives $\qquad \dot{V}_1 = \dot{V}_2 = \dot{V}_3$

or $\qquad D_1^2 u_1 = D_2^2 u_2 + D_3^2 u_3. \qquad \ldots (3)$

From these three equations, u_1, u_2 and u_3 can be calculated and the flow rate determined.

Note that laying two or more pipes in parallel will reduce the frictional resistance and therefore increase the flow rate.

Note also that if the two pipes in parallel are of the same length, diameter and roughness, the flow will be equally divided between them.

Pipeline Including a Pump

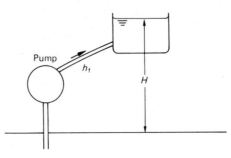

The pump is required to provide the water with sufficient head to increse the potential head by H and overcome the energy degradation h_f.

Head required for the pump = $H + h_f$.
Energy required/unit mass = $g\,(H + h_f)$ (and for a mass flow rate $\rho \dot{V}$).
Power required for the pump = $\rho \dot{V} g\,(H + h_f)$.

If the pump efficiency is η, power required by the pump = $\dfrac{\rho \dot{V} g\,(H + h_f)}{\eta}$.

6.3   Power Transmission Through Pipelines

One of the functions of a pipeline is to transmit power. Normally the working fluid is oil or water (assumed incompressible), or, as in the case of pneumatics, it is air (a compressible fluid). If H is the initial total head (energy) available and h_f is the energy degradation within the pipe due to viscous friction then the available head at exit from the pipe will be $(H - h_f)$. This head may be transmitted as a fluid under pressure, or, by using a suitable nozzle, may be converted into kinetic energy.

The power transmitted will be $\dot{m}g\,(H - h)$ or $\rho \dot{V} g\,(H - h_f)$, and the efficiency of transmission through the pipes is $(H - h_f)/H$

For flow in pipes the head loss due to friction is $h_f = 4fLu^2/2gD$.

Therefore

$$\text{power transmitted} = \rho \dot{V} g \left(H - \frac{4fL}{D}\,\frac{u^2}{2g} \right)$$

and since $\dot{V} = Au$, where A is the pipe area,

$$\dot{W} = \rho A g u \left(H - \frac{4fL}{D}\,\frac{u^2}{2g} \right) = \rho \dot{V} g\,(H - K\dot{V}^2),$$

where $K = \dfrac{64}{2g\pi^2}\,\dfrac{fL}{D^5} = \dfrac{fL}{3.026D^5}$.

With the increase in $\dot{V}$ the available head at exit is reduced, until, when $H = K\dot{V}^2$, the head is zero.

Both when $\dot{V} = 0$ and when $H = K\dot{V}^2$ the power transmitted is zero.

82

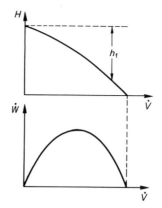

Condition for Maximum Power

Differentiating $\dot{W}$ with respect to $\dot{V}$,

$$\frac{d\dot{W}}{d\dot{V}} = \rho g\,(H - 3K\dot{V}^2).$$

Maximum power when $\quad\dfrac{d\dot{W}}{d\dot{V}} = 0,$

i.e. when $\qquad\qquad H = 3K\dot{V}^2$

when $\qquad\qquad\qquad H = 3h_f \qquad$ (since $h_f = K\dot{V}^2$).

The maximum power transmitted will be

$$\hat{W} = \tfrac{2}{3}\,\rho\dot{V}gH$$

and transmission efficiency $\qquad \eta = \dfrac{H - h_f}{H} = \dfrac{2}{3}.$

6.4 Nozzle at Pipe Outlet

When the power is delivered in the form of a high-velocity jet, a nozzle is required. Assuming there is no energy degradation in the nozzle, the pipe exit head is wholly converted into kinetic energy:

$$H = \frac{4fL}{D}\,\frac{u^2}{2g} = \frac{u_n^2}{g}$$

where u_n = nozzle fluid exit velocity.

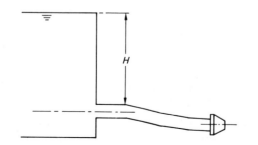

For a given nozzle size, from the mass continuity equation,

$$Au = A_n u_n,$$

where A = pipe cross-sectional area = $\dfrac{\pi D^2}{4}$,

and A_n = nozzle cross-sectional area = $\dfrac{\pi D_n^2}{4}$.

Thus
$$H = \left[\frac{4fL}{D}\left(\frac{D_n}{D}\right)^4 + 1\right]\frac{u_n^2}{2g},$$

from which u_n and the power at nozzle exit can be obtained.

For maximum power from the nozzle, $h_f = \dfrac{H}{3}$

or
$$2 \times \frac{4fL}{D}\frac{u^2}{2g} = \frac{u_n^2}{2g}.$$

i.e.
$$\frac{u_n}{u} = \left(\frac{8fL}{D}\right)^{1/2}$$

and
$$\frac{D}{D_n} = \left(\frac{8fL}{D}\right)^{1/2}$$

Note that if the degradation of energy in the nozzle is significant then we must rewrite

$$u_n = C_u \sqrt{H - \frac{4fL}{D}\frac{u^2}{2g}}$$

giving
$$H = \left[\frac{4fL}{D}\left(\frac{D_n}{D}\right)^4 + \frac{1}{C_u^2}\right]\frac{u_n^2}{2g}.$$

The condition for $\hat{W}$ is also the condition for minimum pipe diameter to transmit a given power. Thus

$$\dot{W} = \rho g \dot{V}\left(\tfrac{2}{3}H\right),$$

giving
$$\dot{V} = \frac{3\dot{W}}{2\rho g H}.$$

Also,
$$h_f = \frac{H}{3} = \frac{4fL}{D}\frac{u^2}{2g} = \frac{fL\,\dot{V}^2}{3.026 D^5}.$$

Hence
$$D^5 = \frac{3}{3.026}\frac{fL\dot{V}^2}{H} = 0.9914\frac{fL\dot{V}^2}{H}.$$

Maximum power transmission when $f = K(Re)^{-n}$:

$$\dot{W} = \dot{m}\frac{u_n^2}{2} = \rho A g C_u^2 \left(Hu - \frac{4fL}{D}\frac{u^3}{2g}\right).$$

Substituting for f,

$$\dot{W} = \rho A g C_u^2 \left[Hu - \frac{4L}{d}K\left(\frac{\mu}{\rho D}\right)^n \frac{u^{3-n}}{2g}\right].$$

for maximum power transmission, $\dfrac{d\dot{W}}{du} = 0.$

i.e.
$$0 = H - \frac{4LK}{D}\left(\frac{\mu}{\rho D}\right)^n (3-n)\frac{u^{3-n}}{g}$$

i.e.

$$0 = H - \frac{4fL}{D} (3 - n) \frac{u^2}{2g},$$

or

$$H = (3 - n) h_f \qquad \text{and} \qquad h_f = \frac{H}{3 - n},$$

$$\dot{W} = \rho A u g C_u^2 \, (H - h_f) = \rho A u g C_u^2 \left(\frac{2 - h}{3 - n} \right) H,$$

where $u = \dfrac{H}{3 - h} = \dfrac{4fL}{D} \dfrac{u^2}{2g}$.

6.5 Friction Factor

The value of f, the friction factor, depends upon the type of flow within the pipe under consideration as discussed earlier. A unique value of f can be found for laminar flow, as shown in Chapter 9 on viscous flow, by using appropriate mathematical reasoning.

For turbulent flow a mathematical derivation of an accurate velocity distribution for pipes is not possible. It is thus necessary to resort to dimensional techniques, intuitive reasoning and experimental procedures to find a value for the friction factor. Flow near the wall shows that a great number of excrescences or protruberances exist and each one of these contributes to the frictional effect. It can be shown that

$$f = \phi \, (\rho, u, D, \mu, \epsilon),$$

where ϵ is a characteristic dimension of the wall roughness, or simply an average height of the excrescences. Dimensional analysis shows that $f = \phi \, [(Re), (\epsilon/D)]$, where ϵ/D is the relative roughness.

A Moody diagram is a plot of f against Reynolds number and ϵ/D and can be found in any standard textbook on fluids. A sketch of the result is given here. (*Note:* (a), (b) and (c) in the diagram refer to values used in example 6.11 later in this chapter.)

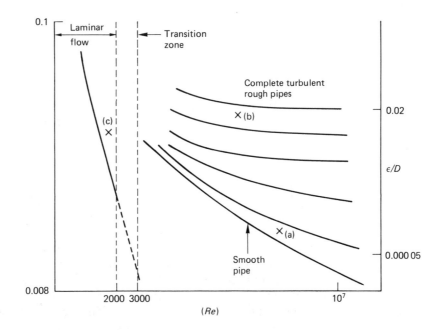

6.6 Worked Examples

Example 6.1

A pipe ABC connecting two reservoirs is of 75 mm diameter. From A to B it is horizontal and from B to C it falls 3.3 m. The lengths of AB and BC are 24 m and 15 m respectively. If the water level in the reservoir at A is 3.7 m above the pipe and the level in the second reservoir is 1 m above the pipe at C, find the quantity which will flow and the absolute pressure head in the pipe at B. Assume f is 0.006 and take the entrance energy degradation as being equal to $\frac{1}{2}u^2/2g$. The water barometer is 10.35 m.

Answer

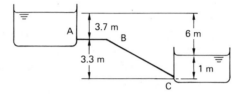

Bernoulli from 1 to 2:

$$\Sigma h_f = 6 \text{ m} = \left(1.5 + \frac{4fL}{D}\right)\frac{u^2}{2g} = \left(1.5 + \frac{4 \times 0.006 \times 39}{0.075}\right)\frac{u^2}{2g},$$

$$\frac{u^2}{2g} = \frac{6}{13.98} \text{ m};$$

or

$$u = \sqrt{\frac{6 \times 2}{13.98} \text{ m} \times 9.81 \frac{\text{m}}{\text{s}^2}} = 2.9 \frac{\text{m}}{\text{s}},$$

$$\dot{V} = Au = \frac{\pi}{4} \times 0.075^2 \text{ m}^2 \times 2.9 \frac{\text{m}}{\text{s}} = 0.0128 \frac{\text{m}^3}{\text{s}}.$$

Bernoulli from 1 to B:

$$\frac{p_1}{\rho g} + \frac{u_1^2}{2g} + z_1 = \frac{p_B}{\rho g} + \frac{u_B^2}{2g} + z_B + \sum_1^B h_f \qquad (u_1 = 0 \text{ in a reservoir}),$$

$$\frac{p_B}{\rho g} = 10.35 \text{ m} + 3.7 \text{ m} - \frac{6}{13.98}\left(1.5 + \frac{4 \times 0.006 \times 24}{0.075}\right) = 10.11 \text{ m}.$$

Example 6.2

Water is discharged from a reservoir through a horizontal pipe 300 mm in diameter for a length of 1600 m, the pipe then suddenly enlarging to 600 mm diameter for a further 1600 m. There are two right-angled easy bends in each length and the difference of static pressure head between the entrance and discharge ends of the pipe is 30.5 m. Calculate the discharge in m^3/s and all energy degradation in the pipe if the value of f is 0.008. Take $h_f = 0.45u^2/2g$ for each bend.

Answer

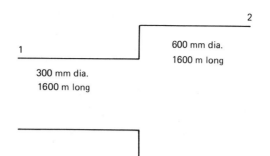

Bernoulli from 1 to 2:

$$\frac{p_1}{\rho g} + \frac{u_1^2}{2g} + z_1 = \frac{p_2}{\rho g} + \frac{u_2^2}{2g} + z_2 + \Sigma h_f.$$

Thus $30.5 \text{ m} = \dfrac{u_2^2}{2g} - \dfrac{u_1^2}{2g} + \dfrac{0.9u_1^2}{2g} + \dfrac{0.9u_2^2}{2g} + \dfrac{(u_1 - u_2)^2}{2g} + \dfrac{4fL_1}{D_1} \times \dfrac{u_1^2}{2g} + \dfrac{4fL_2}{D_2} \times \dfrac{u_2^2}{2g}.$

Continuity: $A_1 u_1 = A_2 u_2$ or $u_1 = 4u_2.$

Thus in above, $30.5 \text{ m} = \dfrac{u_2^2}{2g} [(1 - 16) + 0.9 (1 + 16) + 9 + 2731 + 85.3]$

$$u_2 = \sqrt{\frac{30.5 \times 2 \times 9.81 \ \dfrac{\text{m}^2}{\text{s}^2}}{2825.6}} = 0.46 \ \frac{\text{m}}{\text{s}}$$

and $\qquad \dot{V} = A_2 u_2 = \dfrac{\pi}{4} \times 0.6^2 \ \text{m}^2 \times 0.46 \ \dfrac{\text{m}}{\text{s}} = 0.13 \ \dfrac{\text{m}^3}{\text{s}}.$

Example 6.3

Two reservoirs whose levels differ by 30.5 m are connected by a pipe of diameter 600 mm which is 3050 m long. The pipeline crosses a ridge whose summit is 9.1 m above the level of, and distant 305 m from, the higher reservoir. Find the minimum depth below the ridge at which the pipe must be laid if the absolute head in the pipe is not to fall below 3 m of water, and calculate the discharge in m^3/s ($f = 0.0075$, the water barometer is 10.35 m and the energy degradation at pipe entry and exit is negligible).

Answer

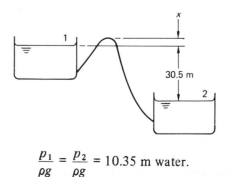

$$\frac{p_1}{\rho g} = \frac{p_2}{\rho g} = 10.35 \text{ m water.}$$

Bernoulli from 1 to 2:

$$z_1 - z_2 = \frac{4fL}{D} \times \frac{u^2}{2g} = 30.5 \text{ m} \qquad \text{(all else is zero)},$$

$$\frac{u^2}{2g} = \frac{30.5 \text{ m} \times 0.6 \text{ m}}{4 \times 0.0075 \times 3050 \text{ m}} = 0.2 \text{ m},$$

$$u = \sqrt{0.4 \times 9.81 \, \frac{\text{m}^2}{\text{s}^2}} = 1.98 \, \frac{\text{m}}{\text{s}},$$

$$\dot{V} = Au = \frac{\pi}{4} \times 0.36 \text{ m}^2 \times 1.98 \, \frac{\text{m}}{\text{s}} = 0.56 \, \frac{\text{m}^3}{\text{s}}.$$

Bernoulli from 1 to 3:

$$\frac{p_1}{\rho g} + \frac{u_1^2}{2g} + z_1 = \frac{p_3}{\rho g} + \frac{u_3^2}{2g} + z_3 + h_{f,13}.$$

$$z_3 - z_1 = x = 10.35 \text{ m} - 3 \text{ m} - \frac{1.98^2}{2g} \text{ m} - \frac{1.98^2}{2g} \left(\frac{4 \times 0.0075 \times 305}{0.6} \right)$$

$$= 7.35 - 0.2 - 3.047 = 4.1 \text{ m}.$$

And depth of pipe below summit = 9.1 − 4.1 = 5 m.

Example 6.4

A pump delivers a flow of 0.0157 m³/s of water from one reservoir to a second whose surface level is 30.5 m above the first. The pipe is of diameter 150 mm for a length of 1.2 km and changes to a diameter of 100 mm for the remaining length of 0.4 km. The resistance coefficient is 0.009. Neglecting any degradation other than that due to pipe friction, find the power required to drive the pump if it has an efficiency of 0.7.

Answer

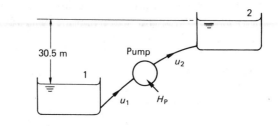

Bernoulli from 1 to 2:

$$H_P = \Sigma h_f + 30.5 \text{ m}$$

$$= 4 \times 0.009 \left[\frac{1200}{0.15} \frac{u_1^2}{2g} + \frac{400}{0.1} \frac{u_2^2}{2g} \right] + 30.5 \text{ m}.$$

Continuity: $A_1 u_1 = A_2 u_2 = \dot{V}.$

Thus $\dfrac{u_1^2}{2g} = \dfrac{\dot{V}^2}{A^2 2g} = \dfrac{0.0157^2 \; \dfrac{m^6}{s^2}}{(\frac{1}{4}\pi \times 0.15^2)^2 \; m^4 \times 2 \times 9.81 \; \dfrac{m}{s^2}} = 0.0402 \; m,$

$$\dfrac{u_2^2}{2g} = \dfrac{0.0157^2}{(\frac{1}{4}\pi \times 0.1^2)^2 \times 2 \times 9.81} = 0.2037 \; m.$$

$H_P = 30.5 + 0.036 \, [321.6 + 814.8] = 71.41 \; m.$

$$\text{Power required} = \dfrac{\rho g \dot{V} \, \Delta p}{\eta} = \dfrac{1000 \; \dfrac{kg}{m^3} \times 9.81 \; \dfrac{m}{s^2} \times 0.0157 \; \dfrac{m^3}{s} \times 71.41 \; m}{0.7}$$

$$\times \left[\dfrac{N \, s^2}{kg \, m}\right] \left[\dfrac{kW \, s}{kN \, m}\right]$$

$$= 15.71 \; kW.$$

Example 6.5

The tubes in an oil cooler are of outside diameter 25 mm and of length 2.5 m, and are so mounted that the centres of any three adjacent tubes form an equilateral triangle of 40 mm side. There are 150 such triangles. Water is pumped between the tubes in a lengthwise direction at 3 m/s.

    Calculate: (a) the water flow rate, (b) the power required to overcome friction if $f = 0.01$.

Answer

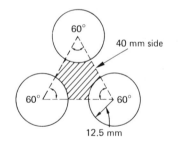

Hydraulic mean diameter:

$$D_e = \dfrac{4A}{\rho}$$

$$= \dfrac{4\left(\dfrac{1}{2} \times 40 \times 40 \; \dfrac{\sqrt{3}}{2} - \dfrac{\pi}{8} \times 25^2\right)}{\pi \times \dfrac{25}{2}}$$

$$= 45.57 \; mm$$

(since shaded area = total triangular area less $3 \times \dfrac{60}{360} \times A_{\text{tube}}$

and perimeter in contact with water = $3 \times \dfrac{60}{360} \times \pi \, D_{\text{tube}}$).

$$h_f = \frac{4fL}{D_e} \times \frac{u^2}{2g} = \frac{4 \times 0.01 \times 2.5 \text{ m}}{0.045\,57 \text{ m}} \times \frac{9 \frac{\text{m}^2}{\text{s}^2}}{2 \times 9.81 \frac{\text{m}}{\text{s}^2}} = 1.0066 \text{ m}.$$

$$\dot{V} = Au = \left(\frac{1}{2} \times 40 \times 40 \frac{\sqrt{3}}{2} - \frac{\pi}{8} \times 25^2 \right) \text{ mm}^2 \left[\frac{\text{m}^2}{10^6 \text{ mm}} \right] \times 3 \frac{\text{m}}{\text{s}} \text{ per triangle}$$

$$= 1.34 \times 10^{-3} \frac{\text{m}^3}{\text{s}}.$$

Total $\dot{V} = 150 \times 1.34 \times 10^{-3} = 0.201 \frac{\text{m}^3}{\text{s}}$ for 150 triangles   (a).

Power required $= \rho g \dot{V} h_f$

$$= 1000 \frac{\text{kg}}{\text{m}^3} \times 9.81 \frac{\text{m}}{\text{s}^2} \times 0.201 \frac{\text{m}^3}{\text{s}} \times 1.0066 \text{ m} \left[\frac{\text{N s}^2}{\text{kg m}} \right] \left[\frac{\text{kW s}}{10^3 \text{ N m}} \right]$$

$$= 1.985 \text{ kW}   \text{(b)}.$$

Example 6.6

A pipe of diameter 1 m is 500 m long and conveys water through a rise of 50 m. If the gauge pressure at the inlet is 750 kN/m² and that at the outlet is 150 kN/m² estimate the flow in the pipe. Consider pipe friction only and take $f = 0.008$. Ignore inlet and outlet velocities.

   If the central portion of the pipe of length 300 m is duplicated, determine the outlet gauge pressure assuming that there is no change in the inlet pressure or in the discharge. Calculate the power available due to the water pressure at the pipe outlet in each case.

$$p_1 = 750 \frac{\text{kN}}{\text{m}^2} \text{ gauge}; \qquad p_2 = 150 \frac{\text{kN}}{\text{m}^2} \text{ gauge}; \qquad f = 0.008.$$

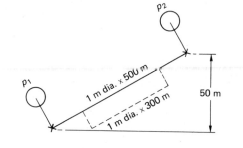

Answer

Bernoulli 1 to 2:

$$\frac{p_1}{\rho g} + \frac{u_1^2}{2g} + z_1 = \frac{p_2}{\rho g} + \frac{u_2^2}{2g} + z_2 + h_f \qquad (u_1 = u_2 = 0),$$

$$h_f = \frac{4fL}{D} \times \frac{u^2}{2g} = \left[\frac{p_1 - p_2}{\rho g} - 50 \right] \text{ m};$$

or
$$\frac{u^2}{2g} = \frac{1 \text{ m}}{4 \times 0.008 \times 500 \text{ m}} \left[\frac{600 \frac{\text{kN}}{\text{m}^2}}{10^3 \frac{\text{kg}}{\text{m}^3} \times 9.81 \frac{\text{m}}{\text{s}^2}} \left[\frac{\text{kg m}}{\text{N s}^2} \right] - 50 \text{ m} \right]$$

$$= 0.698 \text{ m},$$

and
$$u = \sqrt{0.698 \times 2 \times 9.81 \frac{\text{m}^2}{\text{s}^2}} = 3.7 \frac{\text{m}}{\text{s}}.$$

Thus
$$\dot{V} = Au = \frac{\pi}{4} \times 1^2 \text{ m}^2 \times 3.7 \frac{\text{m}}{\text{s}} = 2.91 \frac{\text{m}^3}{\text{s}}.$$

Velocity in the duplicated section $= \frac{u}{2} = 1.85 \frac{\text{m}}{\text{s}}.$

$$\Sigma h_f = \frac{4fL_1}{D_1} \times \frac{u^2}{2g} + \frac{4fL_2}{D_2} \times \frac{u^2}{2g}$$

$$= \frac{4 \times 0.008}{1 \text{ m}} \left[(200 \text{ m} \times 0.7 \text{ m}) + \left(300 \text{ m} \times \frac{0.7}{4} \text{ m} \right) \right] = 6.16 \text{ m}.$$

Bernoulli from 1 to 2:

$$\frac{p_2}{\rho g} = \frac{p_1}{\rho g} - [\Sigma h_f + (z_2 - z_1)]$$

$$= \frac{750 \frac{\text{kN}}{\text{m}^2}}{10^3 \frac{\text{kg}}{\text{m}^3} \times 9.81 \frac{\text{m}}{\text{s}^2}} \left[\frac{\text{kg m}}{\text{N s}^2} \right] - (6.16 + 50) \text{ m} = 20.29 \text{ m},$$

$$p_2 = 20.29 \text{ m} \times 10^3 \frac{\text{kg}}{\text{m}^3} \times 9.81 \frac{\text{m}}{\text{s}^2} \left[\frac{\text{N s}^2}{\text{kg m}} \right] \left[\frac{\text{kN}}{10^3 \text{ N}} \right] = 199.1 \frac{\text{kN}}{\text{m}^2}.$$

Power available due to fluid pressure $= \dot{V} p_2.$

Thus power with single pipe $= 2.91 \frac{\text{m}^3}{\text{s}} \times 150 \frac{\text{kN}}{\text{m}^2} \left[\frac{\text{kW s}}{\text{kN m}} \right] = 436.5 \text{ kW}.$

And power with double pipe $= 2.91 \frac{\text{m}^3}{\text{s}} \times 199.1 \frac{\text{kN}}{\text{m}^2} = 579.4 \text{ kW}.$

Example 6.7

Calculate the power which can be delivered to a machine 7000 m from a hydraulic powerhouse through two horizontal pipes each of diameter 0.1 m laid in parallel. The inlet pressure to the pipes is maintained at 5.6 MN/m² and the efficiency of transmission along the pipes is 90 per cent. If a third pipe of diameter 0.1 m is now used, what decrease in supply pressure can be allowed if the same power at the same pressure is to be maintained at the machine? What is the efficiency of transmission under these conditions?

Assume for all pipes that $f = 0.008$ and neglect all degradation of energy except that due to pipe friction.

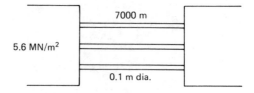

7000 m

5.6 MN/m²

0.1 m dia.

Answer

For an efficiency of transmission of 90 per cent,

$$h_f = 0.1 \frac{p}{\rho g} = \frac{4fL}{D} \frac{u^2}{2g},$$

$$\frac{p}{\rho g} = \frac{5.6 \frac{MN}{m^2}}{10^3 \frac{kg}{m^3} \times 9.81 \frac{m}{s^2}} \left[\frac{kg\,m}{N\,s^2}\right] \left[\frac{10^6\,N}{MN}\right] = 570.8\,m,$$

$$u = \sqrt{\frac{570.8\,m \times 2 \times 9.81 \frac{m}{s^2} \times 0.1\,m \times 0.1}{4 \times 0.008 \times 7000\,m}} = 0.707\,\frac{m}{s}.$$

Power delivered $= \eta p\dot{V} = 0.9 \times 5.6 \times 10^6 \frac{N}{m^2} \times \frac{\pi}{4} \times 0.1^2\,m^2 \times 2 \times 0.707 \frac{m}{s} \left[\frac{kW\,s}{10^3\,N\,m}\right]$

$$= 55.97\,kW.$$

With a third pipe added, $\dot{V}$ is unchanged
and velocity $= \frac{2}{3}$ original velocity.

Thus $$h'_f = \frac{4}{9} h_f$$

Hence $$\Delta p' = \frac{4}{9} \Delta p \qquad \text{where } \Delta p \text{ is pressure drop due to friction.}$$

Thus there is a saving of $\frac{5}{9} \Delta p = \frac{5}{9} \times 0.1 \times 5.6 \times 10^3 \frac{kN}{m^3} = 3.11 \frac{kN}{m^2}$.

Efficiency $\eta' = \dfrac{\text{Final outlet pressure (3 pipes)}}{\text{Initial outlet pressure (2 pipes)}} = \dfrac{5.6 \times 0.9}{5.6 - 0.311} = 0.953$.

Example 6.8

A pipeline of diameter 0.5 m and 2000 m long runs from a reservoir to the nozzle of a Pelton wheel. The nozzle is 300 m below the level of the reservoir. Assuming a coefficient of velocity of 0.98 for the nozzle, and a friction coefficient of 0.009 for the pipe, calculate the diameter of the nozzle for maximum power transmission.

If the overall efficiency of the Pelton wheel, including nozzle, is 0.82, calculate the maximum power output.

For maximum power,

$$\frac{D}{d} = C_u^{1/2} \left(\frac{8fL}{D}\right)^{1/4} = 0.98^{1/2} \left(\frac{8 \times 0.009 \times 2000\,m}{0.5\,m}\right)^{1/4}$$

$$= 4.08\,m,$$

$$d = \frac{D}{4.08} = \frac{0.5}{4.08} = 0.1225\,m.$$

For maximum power, $h_f = \dfrac{H}{3} = \dfrac{300 \text{ m}}{3} = 100 \text{ m} = \dfrac{4fL}{D} \dfrac{u^2}{2g}$.

$$u = \sqrt{\dfrac{100 \text{ m} \times 0.5 \text{ m} \times 2 \times 9.81 \dfrac{\text{m}}{\text{s}^2}}{4 \times 0.009 \times 2000 \text{ m}}} = 3.69 \dfrac{\text{m}}{\text{s}} \; ;$$

$$\hat{W} = \eta_0 \rho \dot{V} g \left(\dfrac{2H}{3}\right)$$

$$= 0.82 \times 10^3 \dfrac{\text{kg}}{\text{m}^3} \times \left(\dfrac{\pi}{4} \times 0.5^2 \text{ m}^2 \times 3.69 \dfrac{\text{m}}{\text{s}}\right) \times 9.81 \dfrac{\text{m}}{\text{s}^2} \times 200 \text{ m} \left[\dfrac{\text{N s}^2}{\text{kg m}}\right] \left[\dfrac{\text{MW s}}{10^6 \text{ N m}}\right]$$

$$= 1.166 \text{ MW}.$$

Example 6.9

(a) If the friction coefficient f in a uniform-bore pipe is given by

$$f = 0.075 \, (Re)^{-1/4},$$

where (Re) = Reynolds number, show that the head loss due to friction is $\frac{4}{11}$ of the supply head, when the power transmitted is a maximum.

(b) A pipe of diameter 150 mm and 3 000 m long is supplied with water at a pressure of 8.3 MN/m². Assuming the kinetic viscosity of water, ν, to be 1.14×10^{-6} m²/s, what is the maximum power that can be transmitted?

Answer

(a) Refer to section 6.4.

$$H = (3 - n)h_f,$$

or $\qquad\qquad H = (3 - \tfrac{1}{4})h_f = \tfrac{11}{4} h_f,$

or $\qquad\qquad h_f = \tfrac{4}{11} H.$

(b) $(Re) = \dfrac{uD}{\nu} = \dfrac{u \times 0.15 \text{ m s}}{1.14 \times 10^{-6} \text{ m}^2} = 131.6 \times 10^3 \, (u) \dfrac{\text{s}}{\text{m}},$

$$f = 0.075 (Re)^{-1/4} = 0.075 \left(131.6 \times 10^3 u \dfrac{\text{s}}{\text{m}}\right)^{-1/4} = 0.003\,94 u^{-1/4} \left(\dfrac{\text{s}}{\text{m}}\right)^{-1/4},$$

$$h_f = \tfrac{4}{11}H = \dfrac{4fL}{D} \dfrac{u^2}{2g},$$

$$\dfrac{4}{11}\left(\dfrac{p}{\rho g}\right) = \dfrac{4fL}{D} \dfrac{u^2}{2g},$$

or $\dfrac{4}{11}\left(\dfrac{8.3 \times 10^6 \dfrac{\text{N}}{\text{m}^2} \text{s}^2}{10^3 \dfrac{\text{kg}}{\text{m}^3} \times 9.81 \text{ m}}\right)\left[\dfrac{\text{kg m}}{\text{N s}^2}\right] = \dfrac{4 \times 0.003\,94 u^{-1/4} \left(\dfrac{\text{s}}{\text{m}}\right)^{-1/4} \times 3000 \text{ m } u^2 \left(\dfrac{\text{m}}{\text{s}}\right)^2}{0.15 \text{ m} \times 2 \times 9.81 \text{ m}}$

or $\qquad\qquad u^{1.75} = 19.15 \left(\dfrac{\text{m}}{\text{s}}\right)^{1.75},$

$$u = 5.4 \text{ m/s}.$$

Power = $\rho g \dot{V} H$,

where $H = \dfrac{7}{11} \left(\dfrac{8.3 \times 10^6 \, \dfrac{N}{m^2} \, s^2}{10^3 \, \dfrac{kg}{m^3} \times 9.81 \, m} \right) \left[\dfrac{kg \, m}{N \, s^2} \right] = 538.4 \, m.$

Power $= 10^3 \, \dfrac{kg}{m^3} \times 9.81 \, \dfrac{m}{s^2} \times \dfrac{\pi}{4} \times 0.15^2 \, m^2 \times 5.4 \, \dfrac{m}{s} \times 538.4 \, m \left[\dfrac{N \, s^2}{kg \, m} \right] \left[\dfrac{kW \, s}{10^3 \, N \, m} \right]$

$= 504 \, kW.$

Example 6.10

The output of a multi-cylinder hydraulic motor is required to be 135 kW, when its efficiency is 73 per cent. A hydraulic power station developing a pressure of 8.3 MN/m² supplies the water through four pipes, each of diameter 75 mm and 3.2 km long. Determine the pressure at the motor, the velocity of flow in the pipes and the efficiency of transmission. Take $f = 0.008$.

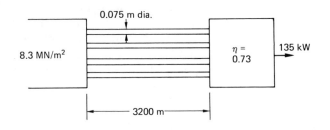

Answer

$$H = \dfrac{p}{\rho g} = \dfrac{8.3 \times 10^6 \, \dfrac{N}{m^2}}{10^3 \, \dfrac{kg}{m^3} \times 9.81 \, \dfrac{m}{s^2}} \left[\dfrac{kg \, m}{N \, s^2} \right] = 846.1 \, m,$$

$$\dot{V} = 4 \times \dfrac{\pi}{4} \, D^2 u = \pi \times 0.075^2 \, m^2 \, u \, \dfrac{m}{s} = 0.0177u \, \dfrac{m^3}{s}.$$

Power supplied to motor $= \rho g \dot{V} (H - h_f).$

$$= 10^3 \, \dfrac{kg}{m^3} \times 9.81 \, \dfrac{m}{s^2} \times 0.0177u \, \dfrac{m^3}{s} \left[846 - \dfrac{4fL}{D} \dfrac{u^2}{2g} \right] m \left[\dfrac{N \, s^2}{kg \, m} \right].$$

Now $\dfrac{4fL}{D} \dfrac{u^2}{2g} = \dfrac{4 \times 0.008 \times 3200 \, m \times u^2 \, \dfrac{m^2}{s^2}}{0.075 \, m \times 2 \times 9.81 \, \dfrac{m}{s^2}} = 69.59 \, u^2 \, m,$

thus $\quad$ Power $= u \, [146.9 - 12.08 \, u^2] \times 10^3 \, \dfrac{N \, m}{s} \left[\dfrac{W \, s}{N \, m} \right],$

thus $\quad$ power $= \dfrac{\text{output of motor}}{\text{efficiency of motor}} = \dfrac{135 \, kW}{0.73} = 184.93 \, kW.$

Thus, equating, $\quad u \, [146.9 - 12.08 \, u^2] = 184.93 \quad\quad$ with u in m/s.

The two solutions for u in this cubic equation are:

$$u = 1.59 \ \frac{m}{s} \qquad \text{and} \qquad u = 2.4 \ \frac{m}{s}.$$

$$\text{Pressure at hydraulic motor} = 8.3 \times 10^6 \ \frac{N}{m^2} - \rho g h_f$$

$$= 8.3 \times 10^6 \ \frac{N}{m^2} - \rho g \ 69.6 u^2.$$

Now, when $u = 1.59$ m/s,

$$\text{pressure} = 8.3 \times 10^6 \ \frac{N}{m^2} - \left(10^3 \ \frac{kg}{m^3} \times 9.81 \ \frac{m}{s^2} \times 69.6 \times 1.59^2 \ m\right)\left[\frac{N \ s^2}{kg \ m}\right]$$

$$= 8.3 \times 10^6 - 1.73 \times 10^6 = 6.57 \times 10^6 \ \frac{N}{m^2} = 6.57 \ \frac{MN}{m^2}.$$

When $u = 2.4 \ \dfrac{m}{s}$, pressure $= (8.3 - 3.93) = 4.37 \ \dfrac{MN}{m^2}$,

$$\eta_{\text{transmission}} = \frac{H - h_f}{H} = \frac{8.3 - 6.57}{8.3} = 0.792,$$

or

$$\eta_{\text{transmission}} = \frac{8.3 - 3.93}{8.3} = 0.526.$$

Example 6.11

Calculate the decrease in fluid pressure in each of the following cases:
(a) air at low pressure flowing at a constant speed of 5 m/s through a 20 m length of galvanised ducting whose cross-section is a square of side 2 m ($\epsilon = 1.5239 \times 10^{-4}$ m);
(b) water flowing at a constant speed of 2 m/s at normal atmospheric temperature through a 20 m length of cast-iron piping whose cross-section is a circle of diameter 0.1 m ($\epsilon = 2.591 \times 10^{-4}$ m);
(c) saturated freon-12 liquid at 300 K flowing at a constant speed of 0.01 m/s through a 20 m length of drawn tubing whose cross-section is a circle of diameter 0.02 m ($\epsilon = 1.5239 \times 10^{-6}$ m).

Take values of μ and ρ from tables by Rogers and Mayhew.* All pipes are horizontal.

Answer

(a) Page 16 of tables gives $\mu = 1.846 \times 10^{-5} \ \dfrac{kg}{m \ s}$,

$$\rho = 1.177 \ \frac{kg}{m^3} \quad \text{at 300 K and low pressure.}$$

For a square of side 2m, $D_m = \dfrac{4A}{p} = \dfrac{4 \times 4 \ m^2}{4 \times 2 \ m} = 2 \ m$

*G. F. C. Rogers and Y. R. Mayhew (1980). *Thermodynamic and Transport Properties of Fluids and Other Data*. Blackwell, Oxford.

and Reynolds number $(Re) = \dfrac{\rho u\, D_m}{\mu} = \dfrac{1.177\ \dfrac{\text{kg}}{\text{m}^3} \times 5\ \dfrac{\text{m}}{\text{s}} \times 2\ \text{m}}{1.846 \times 10^{-5}\ \dfrac{\text{kg}}{\text{m s}}}$

$$= 6.376 \times 10^5, \text{ i.e. fully turbulent flow.}$$

Further, $\qquad \dfrac{\epsilon}{D_m} = \dfrac{1.5239 \times 10^{-4}}{2} = 7.62 \times 10^{-5}.$

From the Moody diagram on page 85 (point (a)), $f = 0.0133$ approx.

Thus $\qquad h_f = \dfrac{4\,fL}{D_m} \times \dfrac{u^2}{2g} = \dfrac{4 \times 0.0133 \times 20\ \text{m}}{2\ \text{m}} \times \dfrac{5^2\ \dfrac{\text{m}^2}{\text{s}^2}}{2 \times 9.81\ \dfrac{\text{m}}{\text{s}^2}} = 0.678$

$$= 0.678\ \text{m}.$$

From Bernoulli's equation, assuming constant u and ρ and neglecting gz changes (horizontal pipe),

$$p_1 - p_2 = \rho g h_f = 1.77\ \dfrac{\text{kg}}{\text{m}^3} \times 9.81\ \dfrac{\text{m}}{\text{s}^2} \times 0.678\ \text{m} \left[\dfrac{\text{N s}^2}{\text{kg m}} \right]$$

$$= 7.827\ \dfrac{\text{N}}{\text{m}^2}. \quad \text{(a)}$$

(b) For water at 20 °C approx., $\mu_f = 1002 \times 10^{-6}\ \dfrac{\text{kg}}{\text{ms}}$ (tables page 10),

$$\rho = 10^3\ \dfrac{\text{kg}}{\text{m}^3}.$$

$$(Re) = \dfrac{\rho u\, D_m}{\mu} \qquad \left(\text{where } D_m = \dfrac{4A}{p} = \dfrac{4\pi D^2}{\pi D} = D = 0.1\ \text{m} \right)$$

$$= \dfrac{10\ \dfrac{\text{kg}}{\text{m}^3} \times 2\ \dfrac{\text{m}}{\text{s}} \times 0.1\ \text{m}}{1002 \times 10^{-6}\ \dfrac{\text{kg}}{\text{m s}}} = 199\,600 \text{ (again fully turbulent).}$$

Also, $\qquad \dfrac{\epsilon}{D_m} = \dfrac{2.591 \times 10^{-4}\ \text{m}}{0.1\ \text{m}} = 2.591 \times 10^{-3}\ \text{m}.$

From the Moody diagram on page 85 (point (b)), $f = 0.056$

and $\quad h_f = \dfrac{4fL}{D_m} \dfrac{u^2}{2g} = \dfrac{4 \times 0.056 \times 20\ \text{m}}{0.1\ \text{m}} \times \dfrac{2^2\ \dfrac{\text{m}^2}{\text{s}^2}}{2 \times 9.81\ \dfrac{\text{m}}{\text{s}^2}}$

$$= 9.134\ \text{m}.$$

Thus $p_1 - p_2 = \rho g h_f = 10\ \dfrac{\text{kg}}{\text{m}^3} \times 9.81\ \dfrac{\text{m}}{\text{s}^2} \times 9.134\ \text{m} \left[\dfrac{\text{N s}^2}{\text{kg m}} \right]$

$$= 89\,600\ \dfrac{\text{N}}{\text{m}^2} \quad (0.896\ \text{bar}). \quad \text{(b)}$$

(c) From page 15, for liquid freon-12 at 300 K, $\mu = 213 \times 10^{-6}\ \dfrac{kg}{ms}$, $\rho = 1304\ \dfrac{kg}{m^3}$.

$$(Re) = \frac{\rho u D_m}{\mu} = \frac{\rho u D}{\mu}\ \text{for a circular pipe}$$

$$= \frac{1304\ \dfrac{kg}{m^3} \times 0.01\ \dfrac{m}{s} \times 0.02\ m}{213 \times 10^{-6}\ \dfrac{kg}{ms}} = 1224\ (\textit{laminar}\ \text{flow})$$

$$\frac{\epsilon}{D_m} = \frac{1.5239 \times 10^{-6}\ m}{0.02\ m} = 7.62 \times 10^{-5} \qquad \text{and} \qquad f = 0.044\ \text{(Moody diagram, page 85, point (c)),}$$

$$h_f = \frac{4fL}{D_m} \times \frac{u^2}{2g} = \frac{4 \times 0.044 \times 20\ m}{0.02\ m} \times \frac{0.01^2\ \dfrac{m^2}{s^2}}{2 \times 9.81\ \dfrac{m}{s^2}}$$

$$= 8.97 \times 10^{-4}\ m.$$

Thus $p_1 - p_2 = \rho g h_f = 1304\ \dfrac{kg}{m^3} \times 9.81\ \dfrac{m}{s} \times \dfrac{8.97}{10^4}\ m\ \left[\dfrac{N\ s^2}{kg\ m}\right]$

$$= 11.475\ \frac{N}{m^2}. \qquad \text{(c)}$$

Exercises

1 Define the pipe friction coefficient f. Explain with reference to a $\log f - \log (Re)$ graph what is meant by laminar flow, smooth turbulent flow, and fully rough turbulent flow in pipes.

Two pipes, each 800 m long and of diameter 150 mm and 75 mm respectively, are joined in series. The difference of overall head between the ends is 30.5 m water and the rate of flow of water through the pipes is 5.4 dm^3/s. Neglecting all degradation of energy other than that due to pipe friction, determine the shear stress at the wall of each portion of the pipeline. Assume f has the same value for each pipe. (*Answer:* $f = 0.00914$, 0.427 N/m^2, 6.83 N/m^2)

2 Two reservoirs having a difference of levels of 25 m are connected by a pipe of diameter 0.3 m and length 8000 m. Flow is to be increased by 15 per cent by using a pipe in parallel over the last 4000 m. Find the diameter of this additional pipe, neglecting all energy degradation except that due to pipe friction and take $f = 0.006$. (*Answer:* 202 mm)

3 A pipe of diameter 150 mm leads from a reservoir to a point 500 m distant where it branches into a pipe 300 m long of diameter 75 mm having an open end and a pipe of diameter 100 mm, 300 m long with a nozzle of diameter 50 mm at the end. Both branch pipes discharge into the atmosphere at a point 18 m below the level of the reservoir.

Calculate the flow in each of the branch pipes and the pressure at the junction if it is 10 m below the level of the reservoir. Take $C_D = 0.96$ for the nozzle, and f for all pipes as 0.01. Neglect energy degradation at entry.
(*Answer:* 0.005 14 m^3/s, 0.0099 m^3/s, 49 kN/m^2)

4 A hosepipe of diameter D and length L is fitted with a nozzle of diameter d at the end. If energy degradation in the nozzle and pipe entry are 10 per cent of

energy degradation in the pipe, show that for a fixed supply head the force of the jet from the nozzle is a maximum when

$$\frac{d}{D} = \left(\frac{D}{4.4fL}\right)^{1/4}.$$

A hosepipe of diameter 0.1 m and 200 m long discharges water from a nozzle fitted to the end of the pipe. The head at entry to the pipe is 40 m measured above the nozzle, which has a coefficient of velocity of 0.97. If the useful energy of the jet is 70 per cent of the supply head, find the flow of water in m^3/s. f for the pipe = 0.009. *(Answer:* 0.013 $m^2/s)$

5 A reservoir feeds a pipe of diameter 200 mm, 300 m long, which branches into two pipes each of diameter 150 mm and each 150 m long. Both branch pipes are fully open at their ends. One branch has outlets all along its length so arranged that half the water entering it is discharged uniformly along its length through these outlets. The far ends of the branch pipes are at the same level, which is 15 m below that of the reservoir. Calculate the discharge from the end of each branch pipe. Disregard all energy degradation except that due to pipe friction and take $f = 0.006$ for all pipes. *(Answer:* 0.0327 m^3/s, 0.0214 $m^3/s)$

7 Dimensional Analysis and Dynamical Similarity

7.1 Introduction

Many problems in fluid mechanics cannot be solved by direct analysis as in the preceding chapters.

It is possible, however, to formulate an alternative method of solution by use of the fact that equations must be dimensionally homogeneous. As a result of this, certain dimensionless ratios which are regularly encountered in fluid mechanics can be defined and can be used for the correlation of experimental data in a convenient form and also in modelling techniques.

The latter are used to simulate actual prototype conditions at reasonable cost. For example, to attempt to discover how a full-sized aircraft behaves in flight by testing in a wind tunnel of full size would be prohibitively expensive. However, its performance can be predicted with accuracy by building a scaled-down model and testing in a small-sized wind tunnel. Measurements on the model can then be related to the predicted performance of the aircraft by the use of dimensionless ratios which are equal for the model and the full-size prototype aircraft.

7.2 Use of Dimensionless Groups

Certain dimensionless groups are of primary significance here, as follows.

Reynolds Number (Re)

This is given by

$$\frac{\rho u D}{\mu},$$

where ρ = fluid density,

u = fluid velocity,

D = a characteristic linear dimension of the system (e.g. a diameter of a cylinder)

and μ = absolute fluid viscosity.

Note that

$$\frac{\rho u D}{\mu} = \frac{\rho u^2}{\dfrac{uu}{D}} = \frac{\text{inertial stress}}{\text{viscous shear stress}}.$$

Mach Number (*Ma*)

This is given by

$$\frac{u}{u_a}, \quad \text{i.e.} \quad \frac{\text{fluid particle velocity}}{\text{acoustic velocity}}.$$

The Mach number is of significance in gas flow.

Froude Number (*Fr*)

In open-channel flow the pressure on the water surface is atmospheric and is not a significant variable and the change in water depth Δz over a length L is the factor affecting the flow pattern.

By dimensional reasoning it can be shown that

$$\frac{g\,\Delta z}{u^2} = \phi\!\left(\frac{u^2}{gL}\right),$$

where $\phi \equiv$ 'a function of'.

Here $g\,\Delta z/u^2$ = head coefficient

and u^2/gL = Froude number, (Fr), $= \dfrac{\rho u^2 L}{\rho g L^2} = \dfrac{\text{inertia force}}{\text{gravity force}}$.

In free-surface flows (no solid surfaces) there is a balance between pressure forces and surface tension forces acting over a given area.

Here
$$\frac{\Delta p}{\rho u^2} = \phi\left(\frac{\rho u^2 L}{\sigma}\right) = \frac{\text{inertia force}}{\text{surface tension}}.$$

When fluid flows past an immersed body, forces are set up owing to pressure changes in the field. In the flow direction we have *drag* forces F_D and perpendicular to the flow we have *lift* forces F_L. Both F_D and F_L depend upon pressure distribution and are not independent, and in this instance

$$\frac{\Delta p}{\rho u^2} = \phi\,[(Re),\,(Ma),\,(Fr)].$$

7.3 Model Testing

As intimated earlier, model testing is used to simulate full-scale fluid flow conditions at a reasonable cost, in order that, by using the same dimensionless groups between the two, full-scale performance can be predicted.

For this to be possible, there must be both geometrical similarity and dynamical similarity between the model and the prototype. Thus, for example, for a cylinder,

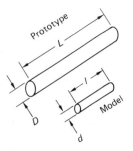

$$\frac{l}{L} = \frac{d}{D} = \lambda$$

for geometrical similarity.
   Again for a square duct:

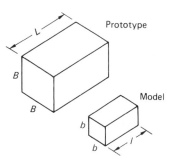

$$\frac{l}{L} = \frac{b}{B} = \lambda$$

for geometrical similarity.
   For dynamical similarity, force, mass and time must be related similarly between model and prototype.

Thus,

force: $F_{\text{model}} = \lambda_a \cdot F_{\text{prototype}}$,
mass: $m_{\text{model}} = \lambda_b \cdot m_{\text{prototype}}$,
time: $t_{\text{model}} = \lambda_c \cdot t_{\text{prototype}}$,

where $\lambda_a = \lambda_b \lambda / \lambda_c^2$ (since force = mass × acceleration).

Buckingham's Pi Method

Buckingham's pi theorem states that for a system of n variables to be correlated with m fundamental dimensions then $(n - m)$ dimensionless groups will result.
   The latter can be developed by use of the following steps:

1. Select the relevant variables in the problem by intuitive reasoning and write all these on one side of the equation.
2. Choose variables which contain all m dimensions in the problem (e.g. length, velocity, density). These should be independent of each other.
3. The dimensionless groups (called Π-groups or parameters) should be written in terms of unknown exponents.
4. The equations relating the exponents for each group should be written and solved.
5. The appropriate dimensionless parameters can be obtained by substitution into the Π-groups.

The procedure is illustrated several times over in the worked examples in this chapter.

Three things should be noted:

(a) If an irrelevant variable has been selected it will appear with an exponent which is zero (as in example 7.1). The efficiency of selection of variables depends upon the degree of experience of the analyst.

(b) The results which accrue from this approach contain certain coefficients whose unknown values can be obtained only by experimental means.

(c) The overriding value of the above approach is that it makes it possible to present a large body of experimental information in a concise manner using non-dimensional groups. For example, a typical performance field for a rotary compressor will give values of ratio of outlet pressure to inlet pressure against non-dimensional mass flow rate for varying values of non-dimensional speed. To adopt any other technique would require the drawing of many graphs to give the same information.

Worked Examples

Example 7.1

The velocity u of a body of mass m falling freely through a distance h is assumed to depend on m and h, on the gravitational constant g and on the initial velocity u_0. Show that the relationship between u and these quantities can be of the form

$$u = \sqrt{gh}\ \phi\ (u_0/\sqrt{gh}).$$

Determine the nature of this function by comparison with the kinematic solution.

Answer

Let
$$u = \phi\ (m, h, g, u_0),$$

i.e.
$$u = m^{a_1} h^{b_1} g^{c_1} u_0^{d_1} + m^{a_2} h^{b_2} g^{c_2} u_0^{d_2} + \ldots \text{etc.}$$

Dividing gives
$$\frac{u}{m^{a_1} h^{b_1} g^{c_1} u_0^{d_1}} = 1 + m^{a_2 - a_1} h^{b_2 - b_1} g^{c_2 - c_1} u_0^{d_2 - d_1} + \ldots \text{etc.}$$

But 1 is a pure number and is dimensionless and thus the l.h.s. is also dimensionless;

i.e.
$$\left[\frac{u}{m^a h^b g^c u_0^d} \right] = 0$$

where [ ] $\equiv$ 'dimensions of',

i.e.
$$u = [m^a h^b g^c u_0^d],$$

and, inserting fundamental dimensions,

$$Lt^{-1} = M^a L^b L^c t^{-2c} L^d t^{-d}.$$

Mass (M)	$0 = a.$	i.e. $a = 0;$	(i.e. mass is *not*
Length (L)	$1 = b + c + d.$	$b = 1 - c - d;$	significant; note
Time (t)	$-1 = -2c - d.$	$c = \dfrac{1-d}{2};$	page 101, §7.4(a))

Thus
$$b = 1 - \left(\frac{1-d}{2} \right) - d = \frac{1}{2} - \frac{d}{2} = \frac{1-d}{2}.$$

Thus
$$u = m^0 h^{(1-d)/2} g^{(1-d)/2} u_0^d + \text{etc. (all other similar terms)}$$

$$= g^{1/2} h^{1/2} \left(\frac{u_0}{g^{1/2} h^{1/2}} \right)^d + \text{etc.} = \sqrt{gh}\ \phi \left(\frac{u_0}{\sqrt{gh}} \right).$$

Now
$$u = \sqrt{u_0^2 + 2gh} = \sqrt{gh}\,\sqrt{2 + \left(\frac{u}{\sqrt{gh}}\right)^2},$$

i.e.
$$\phi(\ ) = \sqrt{2 + \left(\frac{u_0}{\sqrt{gh}}\right)^2}.$$

Example 7.2

It is assumed that the volumetric flow rate through an orifice, $\dot{V}$, is dependent upon ρ and μ, the fluid density and absolute viscosity respectively, d the orifice diameter and p the pressure difference across the orifice. Show that;

$$\dot{V} = d^2 \sqrt{\frac{p}{\rho}}\,\phi\left(\frac{\rho d}{\mu}\sqrt{\frac{p}{\rho}}\right).$$

Hence, putting $p = \rho gh$ where h is the head over the orifice, show that for an orifice in the side of a tank:

$$C_D = \phi(Re).$$

Answer

$$\dot{V} = \phi(\rho, d, p, \mu) = \rho^{a_1} d^{b_1} p^{c_1} \mu^{e_1} + \text{etc.}$$

or
$$\dot{V} = [\rho^a d^b p^c \mu^e],$$

and, inserting dimensions,

$$L^3 t^{-1} = M^a L^{-3a} L^b M^c L^{-c} t^{-2c} M^e L^{-e} t^{-e}.$$

Mass (M) $0 = a + c + e.$

Length (L) $3 = -3a + b - c - e.$

Time (t) $-1 = -2c - e.$

Since μ appears on its own in the above we will get all variables in terms of e.

Thus
$$c = \frac{1 - e}{2}; \qquad a = -e - \frac{1}{2} + \frac{e}{2} = \frac{-e - 1}{2};$$

$$b = 3 + 3\,\frac{-e - 1}{2} + \frac{1 - e}{2} + e = 2 - e;$$

$$\dot{V} = \mu^{(-e-1)/2} d^{2-e} p^{(1-e)/2} \mu^e = d^2 \sqrt{\frac{p}{\mu}} \left(\frac{\rho d}{\mu}\right)^{-e} \left(\frac{p}{\rho}\right)^{-e/2}$$

$$= d^2 \sqrt{\frac{p}{\rho}}\,\phi\left(\frac{\rho d}{\mu}\sqrt{\frac{p}{\rho}}\right).$$

For an orifice, $u^2 = 2gh$ and $\dot{V} = C_D A \sqrt{2gh}$

and if $p = \rho gh,$

$$C_D A \sqrt{2gh} = d^2 \sqrt{gh}\,\phi\left(\frac{\rho d}{\mu}\sqrt{gh}\right),$$

$$C_D\,\frac{\pi d^2}{4}\sqrt{2} = d^2 \phi\left(\frac{\rho u d}{\mu}\sqrt{\frac{gh}{u^2}}\right),$$

or $C_D\,\dfrac{\pi\sqrt{2}}{4} = \phi\left((Re)\sqrt{\frac{1}{2}}\right)$ after substitution

or $C_D = \phi(Re).$

Example 7.3

Show that the flow of liquid over a rectangular notch can be expressed by:

$$\frac{\dot{V}}{L} = g^{1/2} H^{3/2} \phi \left(\frac{\rho g^{1/2} H^{3/2}}{\mu}, \frac{L}{H} \right),$$

where L = notch width,
   H = head over notch,
   ρ = liquid density,
   μ = absolute viscosity
and g = acceleration due to gravity.

It is required to find the flow of liquid of specific gravity 0.8 and viscosity 8 times the viscosity of water over a rectangular notch of width 1 m with a head of 0.6 m by testing a similar notch with water. Determine the corresponding values of H and L for the test notch for dynamical similarity and the scale factor for flow rate.

Answer

$$\frac{V}{L} = \phi (g, H, \rho, L, \mu) = [g^a H^b \rho^c L^d \mu^e]$$

$$L^2 t^{-1} = L^a t^{-2a} L^b M^c L^{-3c} L^d M^e L^{-e} t^{-e}.$$

Mass (M) $0 = c + e.$
Length (L) $2 = a + b - 3c + d - e.$ (Get all variables in terms of e and d).
Time (t) $-1 = -2a - e.$

$$c = -e; \qquad a = \frac{1}{2} - \frac{e}{2};$$

$$b = 2 - \frac{1}{2} + \frac{e}{2} - 3e - d + e = \frac{3}{2} - \frac{3e}{2} - d.$$

$$\frac{\dot{V}}{L} = g^{\frac{1}{2} - \frac{e}{2}} H^{\frac{3}{2} - \frac{3e}{2} - d} \rho^{-e} L^d \mu^e$$

$$= g^{1/2} H^{3/2} \phi \left\{ \left(\frac{\mu}{\rho H^{3/2} g^{1/2}} \right) \cdot \left(\frac{L}{H} \right) \right\} = g^{1/2} H^{3/2} \phi \left\{ \left(\frac{\rho g^{1/2} H^{3/2}}{\mu} \right) \cdot \left(\frac{L}{H} \right) \right\}$$

$$\text{(can invert brackets).}$$

For dynamical similarity (w = water, o = oil),

$$\left(\frac{\rho g^{1/2} H^{3/2}}{\mu} \right)_w = \left(\frac{\rho g^{1/2} H^{3/2}}{\mu} \right)_o \qquad \dots (1)$$

and

$$\left(\frac{L}{H} \right)_w = \left(\frac{L}{H} \right)_o. \qquad \dots (2)$$

In 1, $H_w^{3/2} = \frac{\rho_o g^{1/2} H_o^{3/2}}{\mu_o} \left(\frac{\mu_w}{\rho_w g^{1/2}} \right) = \frac{0.8}{1} \times \frac{1}{8} \times 0.6^{3/2}$ (or $H_w = 0.129$ m).

In 2, $L_w = L_o \left(\frac{H_w}{H_o} \right) = 1 \text{ m} \times \frac{0.129}{0.6} = 0.215$ m.

Now $$\left(\frac{\dot{V}/L}{g^{1/2} H^{3/2}} \right)_w = \left(\frac{\dot{V}/L}{g^{1/2} H^{3/2}} \right)_o;$$

Thus $$\frac{\dot{V}_o}{\dot{V}_w} = \frac{L_o}{L_w} \left(\frac{H_o}{H_w} \right)^{3/2} = \frac{1}{0.215} \left(\frac{0.6}{0.129} \right)^{3/2} = 46.7.$$

Example 7.4

Show that the torque required to rotate a disc of diameter d at an angular velocity ω in a fluid of density ρ, viscosity μ, is given by

$$\tau = d^5 \omega^2 \rho \, \phi \left(\frac{\rho d^2 \omega}{\mu} \right).$$

A disc of diameter 230 mm absorbs 160 W when rotated in water at a speed of 146 rad/s. What will be the corresponding speed of rotation of a similar disc of diameter 690 mm when it rotates under dynamically similar conditions in air? Calculate the power absorbed at this speed. Take $\mu_w = 101.3 \times 10^{-5}$ N s/m^2, $\rho_a = 1.25$ kg/m^3, $\mu_a = 1.85 \times 10^{-5}$ N s/m^2.

Answer

$$\tau = \phi \, (d, \, \omega, \, \rho, \, \mu) = [d^a \omega^b \rho^c \mu^e].$$

$$\mathrm{ML}^2 \mathrm{t}^{-2} = \mathrm{L}^a \mathrm{t}^{-b} \mathrm{M}^c \mathrm{L}^{-3c} \mathrm{M}^e \mathrm{L}^{-e} \mathrm{t}^{-e}.$$

Mass (M) $1 = c + e.$
Length (L) $2 = a - 3c - e.$ Get all in terms of e.
Time (t) $-2 = - b - e.$

$$b = 2 - e; \qquad c = 1 - e; \qquad a = 2 + 3 - 3e + e = 5 - 2e.$$

$$\tau = d^{5-2e} \omega^{2-e} \rho^{1-e} \mu^e = d^5 \omega^2 \phi \left(\frac{\rho d^2 \omega}{\mu} \right).$$

For dynamical similarity, $\quad \left(\dfrac{\rho d^2 \omega}{\mu} \right)_w = \left(\dfrac{\rho d^2 \omega}{\mu} \right)_a,$

and thus $\quad \omega_a = \dfrac{\rho_w d_w^2 \omega_w \mu_a}{\mu_w \rho_a d_a^2} = \dfrac{10^3}{1.25} \times \dfrac{1.85 \times 10^{-5}}{101.3 \times 10^{-5}} \left(\dfrac{230}{690} \right)^2 \times 146 \, \dfrac{\mathrm{rad}}{\mathrm{s}}$

$$= 237 \, \frac{\mathrm{rad}}{\mathrm{s}}.$$

$$\tau_w = \frac{\dot{W}_w}{\omega_w} = \frac{160 \text{ W}}{146 \, \dfrac{\mathrm{rad}}{\mathrm{s}}} = 1.096 \text{ N m} \qquad \left(1 \text{ W} = 1 \, \frac{\text{N m}}{\text{s}} \right).$$

Thus $\quad \tau_a = \tau_w \left(\dfrac{d_a}{d_w} \right)^5 \left(\dfrac{\omega_a}{\omega_w} \right)^2 \dfrac{\rho_a}{\rho_w} = \tau_w \left(\dfrac{690}{230} \right)^5 \left(\dfrac{237}{146} \right)^2 \times \dfrac{1.25}{1000} = 0.8 \tau_w$

$$= 0.876 \text{ N m},$$

$$\dot{W}_a = \tau_a \cdot \omega_a = 0.876 \text{ N m} \times 237 \, \frac{1}{\text{s}} = 208 \text{ W}.$$

Example 7.5

Show by the method of dimensional analysis that, for a hydrostatic journal bearing, the relationship between the pressure of the oil at the inlet port p, the inlet port diameter d, housing diameter D, mean radial clearance h, mean velocity u through the inlet port, viscosity μ and density ρ of the oil may be expressed as:

$$p = \frac{\mu u}{d} \, \phi \left(\frac{\rho u d}{\mu}, \frac{D}{d}, \frac{h}{d} \right).$$

In a test carried out on a hydrostatic journal bearing it was found that the inlet port pressure was 21 kN/m² when the oil flow rate was 15 litre/min. The oil used in the test had a viscosity of 0.15 P and a density of 950 kg/m³. The inlet port diameter was 50 mm, the housing diameter 0.5 m and the mean radial clearance 0.5 mm. Calculate the inlet port diameter, radial clearance, oil flow rate and inlet port pressure to be expected in a geometrically similar bearing which has a housing diameter of 100 mm if the oil to be used has a viscosity of 0.2 P and a density of 900 kg/m³.

Answer

$$p = \phi(d, D, h, u, \mu, \rho) = d^{a_1} D^{a_2} h^{a_3} u^{a_4} \mu^{a_5} \rho^{a_6}$$

$$\frac{ML}{t^2 L^2} = L^{a_1} L^{a_2} L^{a_3} \left(\frac{L}{t}\right)^{a_4} \left(\frac{M}{Lt}\right)^{a_5} \left(\frac{M}{L^3}\right)^{a_6}.$$

Mass (M) $1 = a_5 + a_6.$
Length (L) $-1 = a_1 + a_2 + a_3 + a_4 - a_5 - 3a_6.$
Time (t) $-2 = -a_4 - a_5.$

Collect in terms of a_2, a_3 and a_5:

$$a_6 = 1 - a_5; \qquad a_1 = 3a_6 + a_5 - a_4 - a_3 - a_2 - 1$$
$$a_4 = 2 - a_5; \qquad\quad = 3 - 3a_5 + a_5 - 2 + a_5 - a_3 - a_2 - 1$$
$$= -a_5 - a_3 - a_2.$$

$$p = d^{-a_5 - a_3 - a_2} D^{a_2} h^{a_3} u^{2-a_5} \mu^{a_5} \rho^{1-a_5}$$

$$= \frac{\mu u}{d} d^{-a_5 - a_3 - a_2 + 1} D^{a_2} h^{a_3} u^{1-a_5} \mu^{a_5 - 1} \rho^{1-a_5}$$

$$= \frac{\mu u}{d} \phi \left\{ \left(\frac{\rho u d}{\mu}\right), \left(\frac{D}{d}\right), \left(\frac{h}{d}\right) \right\}.$$

$$d_2 = D_2 \frac{d_1}{D_1} = 50 \text{ mm} \times \frac{100}{500} = 10 \text{ mm}.$$

$$h_2 = h_1 \frac{d_2}{d_1} = 0.5 \text{ mm} \times \frac{10}{50} = 0.1 \text{ mm}.$$

$$u_1 = \frac{\dot{V}_1}{(\pi D h)_1} = \frac{15 \frac{\text{litre}}{\text{min}} \left[\frac{m^3}{10^3 \text{ litre}}\right]}{\pi \times 0.5 \text{ m} \times 0.0005 \text{ m}} = \frac{60}{\pi} \frac{\text{m}}{\text{min}},$$

$$u_2 = \left(\frac{\rho u d}{\mu}\right)_1 \left(\frac{\mu}{\rho d}\right)_2 = \frac{950}{900} \times \frac{50}{10} \times \frac{0.2}{0.15} \times \frac{60}{\pi} \frac{\text{m}}{\text{min}} = \frac{3800}{9\pi} \frac{\text{m}}{\text{min}}.$$

$$\dot{V}_2 = \pi_2 D_2 h_2 u_2 = \pi \times 0.1 \text{ m} \times 0.0001 \text{ m} \times \frac{3800}{9\pi} \frac{\text{m}}{\text{min}} \left[\frac{10^3 \text{ litre}}{m^3}\right]$$

$$= 4.22 \frac{\text{litre}}{\text{min}}.$$

$$p_2 = p_1 \frac{\rho_2}{\rho_1} \frac{u_2^2}{u_1^2} = 21 \frac{\text{kN}}{\text{m}^2} \times \frac{900}{950} \times \frac{3800^2}{(9\pi)^2} \times \frac{\pi^2}{60^2}$$

$$= 985 \frac{\text{kN}}{\text{m}^2}.$$

Example 7.6

Show by the method of dimensional analysis that for a screw propeller the relation between the thrust F, torque τ, diameter D, speed of advance u, speed of rotation N and density and viscosity of the fluid ρ and μ may be put in the form

$$F = \rho D^2 u^2 \phi\left(\frac{\rho D^3 u^2}{\tau}, \frac{ND}{u}, \frac{\rho u D}{\mu}\right).$$

A propeller of diameter 3 m is to run at 6.5 m/s and its performance is to be estimated by means of a similar model of diameter 0.3 m. Results obtained from the model were:

speed of advance: 2.25 m/s at 505 rev/min;
torque: 6 N m;
thrust: 100 N.

Determine corresponding characteristics of the prototype and calculate its efficiency, assuming that the effect of the Reynolds number is negligible.

Answer

$$F = \phi(\tau, D, u, N, \rho, \mu) = \tau^a D^b u^c N^d \rho^e \mu^f.$$

$$\left(\frac{ML}{t^2}\right) = \left(\frac{ML^2}{t^2}\right)^a L^b \frac{L^c}{t^c} \frac{1}{t^d} \left(\frac{M}{L^3}\right)^e \left(\frac{M}{Lt}\right)^f.$$

Mass (M) $1 = a + e + f.$
Length (L) $1 = 2a + b + c - 3e - f.$ } Get a, b and c in terms of d, e and f.
Time (t) $-2 = -2a - c - d - f.$

$a = 1 - e - f;$
$c = 2 - 2a - d - f = 2 - 2(1 - e - f) - d - f = 2e - d + f;$
$b = 1 - 2a - c + 3e + f = 1 - 2(1 - e - f) - (2e - d + f) + 3e + f;$
$b = -1 + d + 2f + 3e.$

Thus $F = \tau^{1-e-f} D^{-1+d+2f+3e} u^{2e+d-f} N^d \rho^e \mu^f$

$= \rho D^2 u^2 [\tau^{1-e-f} D^{-3+d+2f+3e} u^{2e-d+f-2} N^d \rho^{e-1} \mu^f]$

$= \rho D^2 u^2 \left[\left(\frac{\rho D^3 u^2}{\tau}\right)^{f+e-1} \rho^{-f} D^{-f+d} u^{-f-d} N^d \mu^f\right]$

$= \rho D^2 u^2 \left[\left(\frac{\rho D^3 u^2}{\tau}\right)^{f+e-1} \left(\frac{\rho u D}{\mu}\right)^{-f} \left(\frac{ND}{u}\right)^d\right]$

$= \rho D^2 u^2 \phi\left[\left(\frac{\rho D^3 u^2}{\tau}\right), \left(\frac{\rho u D}{\mu}\right), \left(\frac{ND}{u}\right)\right].$

For dynamical similarity, $N_p = N_m \times \dfrac{D_m}{D_p} \times \dfrac{u_p}{u_m}$ (p = prototype, m = model)

$$= 505 \frac{\text{rev}}{\text{min}} \times \frac{0.3}{3} \times \frac{6.5}{2.25} = 145.9 \frac{\text{rev}}{\text{min}}.$$

$$\tau_p = \tau_m \frac{\rho_p}{\rho_m} \frac{D_p^3}{D_m^3} \frac{u_p^2}{u_m^2} = 6 \text{ N m} \times \frac{1}{1} \times \left(\frac{3}{0.3}\right)^3 \times \left(\frac{6.5}{2.25}\right)^2 = 500\,74 \text{ N m},$$

$$F_p = F_m \frac{\rho_p D_p^3 u_p^2}{\rho_m D_m^3 u_m^2} = 100 \text{ N} \times \frac{1}{1} \times \left(\frac{3}{0.3}\right)^2 \times \left(\frac{6.5}{2.25}\right)^2 = 83\,457 \text{ N}.$$

$$\text{Efficiency} = \frac{F_p u_p}{\tau_p \omega_p} = \frac{83\,457\text{ N} \times 6.5\,\frac{\text{m}}{\text{s}}}{50\,074\text{ N m} \times 145.9\,\frac{\text{rev}}{\text{min}}\left[\frac{\text{min}}{60\text{ s}}\right]\left[\frac{2\pi\text{ rad}}{\text{rev}}\right]} = 0.709.$$

(*Note:* ρ is the same for both prototype and model.)

Example 7.7

The pressure rise Δp generated by a pump depends on the impeller diameter D, its rotational speed N, the fluid density ρ and viscosity μ and the rate of discharge $\dot{V}$. Show that the relationship between these variables may be expressed as:

$$\Delta p = \rho N^2 D^2 \phi \left[\left(\frac{\dot{V}}{ND^3}\right), \left(\frac{\rho ND^2}{\mu}\right) \right].$$

A given pump rotates at a speed of 1000 rev/min, and at its duty point it generates a head of 12 m when pumping water at a rate of 15 litre/s. Calculate the head generated by a similar pump, twice the size, when operating under dynamically similar conditions and discharging 45 litre/s. Calculate also the shaft power of the second pump if its overall efficiency is 80 per cent. The influence of the Reynolds number is negligible.

Answer

$$\Delta p = \phi\,(D, N, \rho, \mu, \dot{V}) = D^a N^b \rho^c \mu^d \dot{V}^e.$$

$$\frac{M}{Lt^2} = L^a\,\frac{1}{t^b}\,\frac{M^c}{L^{3c}}\,\frac{M^d}{L^d t^d}\,\frac{L^{3e}}{t^e}.$$

Mass (M) $\quad 1 = c + d.$
Length (L) $\quad -1 = a - 3c - d + 3e.$ $\Big\}$ $\quad$ Get a, b and c in terms of d and e.
Time (t) $\quad -2 = -b - d + e.$

$c = 1 - d;$ $\qquad a = 3c + d - 3e - 1 = 3(1 - d) + d - 3e - 1 = 2 - 2d - 3e;$
$b = 2 - d - e;$

$$\Delta p = D^{2-2d-3e}\,N^{2-d-e}\,\rho^{1-d}\,\mu^d\,\dot{V}^e$$

$$= \rho N^2 D^2 \left[\left(\frac{\dot{V}}{ND^3}\right)^e \left(\frac{\rho ND^2}{\mu}\right)^{-d} \right]$$

$$= \rho N^2 D^2 \phi \left[\left(\frac{\dot{V}}{ND^3}\right), \left(\frac{\rho ND^2}{\mu}\right) \right]$$

Note: $\dfrac{\rho ND^2}{\mu} \equiv \dfrac{\rho \omega D^2}{\mu} \equiv \dfrac{\rho \omega \frac{D}{2} D}{\mu} \equiv \dfrac{\rho u D}{\mu} \equiv (Re)$ $\qquad$ dimensionally.

For dynamical similarity, $\qquad \left(\dfrac{\dot{V}}{ND^3}\right)_m = \left(\dfrac{\dot{V}}{ND^3}\right)_P.$

Thus $\quad N_P = N_m \times \dfrac{\dot{V}_P}{\dot{V}_m} \times \dfrac{D_m^3}{D_P^3} = 1000\,\dfrac{\text{rev}}{\text{min}} \times \dfrac{45}{15} \times \left(\dfrac{1}{2}\right)^3 = 375\,\dfrac{\text{rev}}{\text{min}}.$

$$\Delta p_P = \Delta p_m \frac{\rho_P N_P^2 D_P^2}{\rho_m N_m^2 D_m^2} \qquad \text{and} \qquad \Delta p = \rho g\,\Delta H,$$

i.e. $\Delta H_P = \Delta H_m \dfrac{\rho_P N_P^2 D_P^2}{\rho_m N_m^2 D_m^2} = 12\text{ m} \times \dfrac{1}{1}\left(\dfrac{375}{1000}\right)^2 \left(\dfrac{2}{1}\right)^2 = 6.75\text{ m.}$

Shaft power $= (\rho \dot{V} g \, \Delta H)/\eta$

$$= \frac{10^3 \; \frac{\text{kg}}{\text{m}^3} \times 45 \; \frac{\text{litre}}{\text{s}} \left[\frac{\text{m}^3}{10^3 \; \text{litre}}\right] \times 9.81 \; \frac{\text{m}}{\text{s}^2} \times 6.75 \; \text{m}}{0.8} \left[\frac{\text{N s}^2}{\text{kg m}}\right] \left[\frac{\text{kW s}}{10^3 \; \text{N m}}\right]$$

$= 3.725 \; \text{kW}.$

Exercises

1 Show, by dimensional analysis, that the resistance F of a sphere of diameter D moving with a constant velocity u in a fluid of density ρ and viscosity μ may be expressed by

$$F = \frac{\mu^2}{\rho} \; \phi\left(\frac{\rho u D}{\mu}\right),$$

where ϕ is 'a function of'. It was observed that a steel ball of diameter 2.5 mm and density 7750 kg/m³ falls with a uniform velocity of 5.0 mm/s in a liquid of density 83 kg/m³. Given that the above function ϕ is such that $\phi(z) = 10z$ for any z, determine the viscosity of the liquid. *(Answer:* 4.94 N s/m²)

2 Show that for a pipe of diameter d the pressure loss per unit length due to friction for an incompressible fluid of density ρ, viscosity μ and velocity u is given by:

$$\frac{p}{L} = \frac{\rho u^2}{d} \; \phi\,(Re).$$

The friction loss in a 50 mm diameter pipe carrying air is to be estimated from a suitable test on a similar pipe 100 mm in diameter carrying water. Determine the water velocity required to give dynamical similarity for an air velocity of 12 m/s. What is the ratio of pressure drops per unit length water to air?

(Answer: 0.411 m/s, 0.47)

3 It is judged that the performance of a lubricating oil ring depends upon the following variables: quantity of oil delivered per unit time $\dot{V}$, inside diameter of the ring D, shaft speed in revolutions per unit time N, oil viscosity μ, oil density ρ, specific weight of the oil w ($= \rho g$) and surface tension in air σ. Show that

$$\frac{\dot{V}}{D^3 N} = \phi\left(\frac{\rho N D^2}{\mu}, \frac{\mu N D}{\sigma}, \frac{w D}{\mu N}\right).$$

8 Open-channel Flow

8.1 Introduction

An open channel is one in which there is a free surface at atmospheric pressure e.g. canals, rivers and pipes which are not running full.

Turbulent uniform flow will occur in channels of constant cross-section when the flow has become fully established.

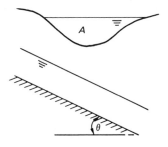

Example

Consider an open channel of cross-sectional area A, wetted perimeter P and slope θ.

Since flow is uniform along the channel and the pressure remains constant with distance, the degradation of mechanical energy due to friction will be equal to the loss of potential energy, i.e. the head loss due to the slope of the channel. Hence the channel gradient is equal to the hydraulic gradient.

$$\sin \theta = i = \frac{h_f}{L}.$$

But

$$h_f = f\, \frac{A_s}{A}\, \frac{u^2}{2g},$$

where A_s = surface area = PL,

$$h_f = f\, \frac{PL}{A}\, \frac{u^2}{2g}$$

and

$$i = \frac{h_f}{L} = f\, \frac{P}{A}\, \frac{u^2}{2g}.$$

Writing $\dfrac{A}{P} = R_m$, the hydraulic mean radius ($= \tfrac{1}{4} D_m$),

$$i = \frac{f}{R_m}\, \frac{u^2}{2g},$$

or $\quad u = \sqrt{\dfrac{2g}{f} \; R_m i} = C\sqrt{R_m i}$ $\qquad$ (Chezy formula) $\left(C = \sqrt{\dfrac{2g}{f}}\right)$.

C is the Chezy constant ($m^{1/2}/s$).

The value of C will vary with surface conditions of the channel. The smoother the surface the higher will be the value of C.

8.2 Manning Formula

As a result of analysing experimental data from open channels, Manning summarised the results with an empirical expression of the form

$$u = MR_m^{2/3} i^{1/2},\qquad \text{where } M \text{ is a constant,}$$

depending on channel roughness.

i.e. $$C = MR_m^{1/6}.$$

Discharge along an open channel is

$$\dot{V} = Au = AC\sqrt{R_m i} = AMR_m^{2/3} i^{1/2}.$$

8.3 Flow-measuring Devices for Open-channel Flow—Notches and Weirs

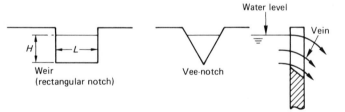

A notch may be regarded as an orifice with the water surface below the upper edge. It is generally rectangular or triangular (vee-notch).

A weir is a wall across a stream over which the water flows. It may be sharp-crested or broad-crested.

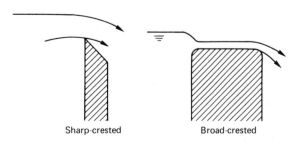

Notches and weirs shorter than the channel width will cause end contractions in the water flowing over them. Suppressed weirs span the channel and cause no end contractions.

For small flow-rates the vee-notch is most accurate; for large flow-rates rectangular weirs are used.

Analysis Based on Bernoulli's Equation

Rectangular Notch or Weir

Consider flow along the streamline shown; upstream at point 1 the velocity is assumed to be negligible; at point 2 in the vena contracta streamlines are assumed to be parallel, i.e. no variation in pressure across the vein and hence the pressure is atmospheric.

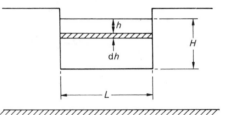 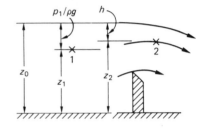

Applying Bernoulli between points 1 and 2,

$$\frac{p_1}{\rho g} + \frac{u_1^2}{2g} + z_1 = \frac{p_2}{\rho g} + \frac{u_2^2}{2g} + z_2 \qquad (u_1 = p_2 = 0).$$

But
$$\frac{p_1}{\rho g} + z_1 = z_0$$

and
$$z_0 - z_2 = h.$$

Thus
$$h = \frac{u_2^2}{2g}$$

or
$$u_2 = \sqrt{2gh}.$$

The discharge, $d\dot{V}$, through a strip of length L and thickness dh at point 2 is given by

$$d\dot{V} = L \, dh \, \sqrt{2gh}.$$

The total discharge over the notch is given by

$$\dot{V} = \int_0^H L\sqrt{2gh} \, dh = \tfrac{2}{3}L\sqrt{2g} \, H^{3/2}.$$

Allowing for contraction of the vein (in the vertical plane) and internal friction due to viscosity, the actual discharge is less and a correction coefficient, the coefficient of discharge, is introduced:

$$\dot{V} = \tfrac{2}{3} C_D L \sqrt{2g} \, H^{3/2}.$$

112

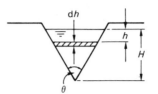

Consider the discharge through a narrow strip of thickness d*h* and depth *h*. The breadth will equal

$$2(H - h)\tan\theta/2.$$

Applying Bernoulli's equation as before, the velocity at the notch is given by

$$u = \sqrt{2gh}.$$

The discharge through the strip is

$$\mathrm{d}\dot{V} = 2(H - h)\tan\frac{\theta}{2}\sqrt{2gh}\ \mathrm{d}h.$$

Total discharge through the notch is

$$\dot{V} = \int_0^H 2\tan\frac{\theta}{2}\sqrt{2g}(H - h)h^{1/2}\ \mathrm{d}h$$

or

$$\dot{V} = 2\tan\frac{\theta}{2}\sqrt{2g}\left[\frac{2}{3}Hh^{3/2} - \frac{2}{5}h^{5/2}\right]_0^H$$

$$= \frac{8}{15}\tan\frac{\theta}{2}\sqrt{2g}H^{5/2}.$$

Allowing for the contraction of the vein and energy degradation,

$$\dot{V} = \frac{8}{15}C_\mathrm{D}\tan\frac{\theta}{2}\sqrt{2g}H^{5/2}. \qquad\qquad \ldots(2)$$

Effect of End Contractions on the Rectangular Weir

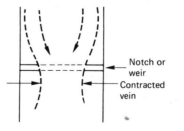

In channels wider than the notch or weir, the vein will contract (in the horizontal plane) after passing over the notch. Francis found experimentally that the contraction was proportional to the head and the discharge was given by:

$$\dot{V} = 1.84 \left(1 - 0.1n\,\frac{H}{L}\right) LH^{3/2}, \qquad \qquad \ldots(3)$$

where n is the number of contractions. Compared with the preceding analysis this gives a value of $C_D = 0.623$. Francis's formula is an empirical one in which the constant in SI units is $1.84\ m^{1/2}/s$.

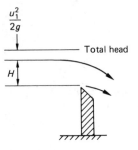

The fluid in the approach channel possesses kinetic energy (so far assumed negligible). Thus the static head measured in the approach channel is less than the total head by $u_1^2/2g$.

Total head $= H + u_1^2/2g$ where, if the channel area is A, $u_1 = \dot{V}/A$.

Substituting in the original equation for a rectangular notch or weir gives

$$\dot{V} = \int_{u_1^2/2g}^{H + u_1^2/2g} L\sqrt{2gh}\ \mathrm{d}h = \frac{2}{3}L\sqrt{2g}\left[\left(H + \frac{u_1^2}{2g}\right)^{3/2} - \left(\frac{u_1^2}{2g}\right)^{3/2}\right].$$

Allowing for friction and contraction of vein and neglecting $u_1^2/2g$ compared with $H + u_1^2/2g$,

$$\dot{V} = \frac{2}{3}C_D L\sqrt{2g}\left(H + \frac{u_1^2}{2g}\right)^{3/2}.$$

Approximate Method of Discharge Calculation

(a) Assume H = the total head over the weir and calculate $\dot{V}$ from either the Francis formula (3) or equation 1.

(b) Knowing the channel cross-sectional area A, calculate

$$u_1\ (= \dot{V}/A) \qquad \text{and} \qquad u_1^2/2g.$$

(c) Obtain $(H + u_1^2/2g)$ and use this to obtain a second approximation of $\dot{V}$.

This value is usually accurate enough, although the process could be repeated.

8.4 Worked Examples

Example 8.1

A channel of trapezoidal section, with sides sloping at $45°$, conveys water to a depth of 0.75 m. Find the width of the base and the channel gradient to discharge $1.2\ m^3/s$ with a mean velocity of 0.8 m/s. Take the Chezy constant as $66\ m^{1/2}/s$.

Answer

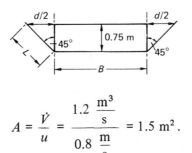

$$A = \frac{\dot{V}}{u} = \frac{1.2 \ \frac{m^3}{s}}{0.8 \ \frac{m}{s}} = 1.5 \ m^2.$$

Cross-sectional area $= (0.75B) + \left(\frac{d}{2} \times 0.75\right) = 1.5 \ m^2.$

Now $\qquad\qquad \frac{d}{2} = 0.75 \tan 45° = 0.75 \ m$

and $\qquad\qquad L = 0.75 \sqrt{2} = 1.061 \ m.$

Thus $\qquad\qquad B = \frac{1.5 - 0.75^2}{0.75} = 1.25 \ m.$

Hydraulic mean radius $= R_m = \frac{A}{p} = \frac{1.5 \ m^2}{[1.25 + (2 \times 1.061)] \ m} = 0.445 \ m$

Thus slope $i = \frac{u^2}{C^2 R_m} = \frac{0.8^2 \ \frac{m^2}{s^2}}{\left(66 \ \frac{m^{1/2}}{s}\right)^2 \times 0.445 \ m\right)} = \frac{1}{3029}.$

Example 8.2

Determine the radius of an open channel of semi-circular cross-section which is required to convey water with a flow of 0.3 m³/s, when running full. The slope of the channel is 1 in 2500. Take the Chezy constant as 55 m$^{1/2}$/s.

If the channel was rectangular in form, but of the same total width and depth, what would be the flow when running full, with the same value of Chezy constant and the same slope?

Answer

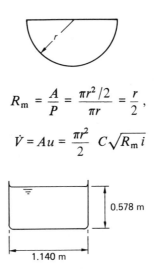

$$R_m = \frac{A}{P} = \frac{\pi r^2/2}{\pi r} = \frac{r}{2},$$

$$\dot{V} = Au = \frac{\pi r^2}{2} \ C\sqrt{R_m i}$$

115

$$= 0.3 \; \frac{m^3}{s} = \frac{\pi}{2} \times 55 \; \frac{m^{1/2}}{s} \sqrt{\frac{1}{2 \times 2500}} \, r^{5/2}$$

$$r^{5/2} = \frac{0.3}{1.222} = 0.2455,$$

$$r = 0.2455^{2/5} = 0.570 \text{ m.}$$

$$R_m = \frac{A}{P} = \frac{0.57 \times 1.140}{1.140 + 2\,(0.57)} = 0.285 \text{ m.}$$

$$\dot{V} = AC\sqrt{R_m \, i} = 0.57 \text{ m} \times 1.14 \text{ m} \times 55 \; \frac{m^{1/2}}{s} \sqrt{\frac{0.285 \text{ m}}{2500}}$$

$$= 0.382 \; \frac{m^3}{s} \, .$$

Example 8.3

Water flows along a channel of circular section of diameter 0.6 m at the rate of 0.15 m³/s. The slope of the channel is 1 in 2000 and the depth of water is 0.45 m. Calculate the value of the Chezy constant.

Answer

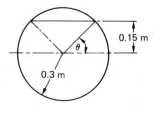

$$\theta = \arcsin \frac{0.15}{0.3} = 30°.$$

$$\text{Flow area} = \pi \times 0.3^2 \times \frac{240}{360} + 0.3 \cos 30° \times 0.15$$

$$= 0.2275 \text{ m}^2 .$$

$$P = 2\pi \times 0.3 \times \frac{240}{360} = 1.257 \text{ m.}$$

$$R_m = \frac{A}{P} = \frac{0.2275}{1.257} = 0.181 \text{ m.}$$

$$u = \frac{\dot{V}}{A} = \frac{0.15 \; \frac{m^3}{s}}{0.2275 \text{ m}^2} = 0.659 \; \frac{m}{s} = C\sqrt{R_m \, i}.$$

Thus
$$C = \frac{0.659 \frac{m}{s} \sqrt{2000}}{\sqrt{0.181} \text{ m}^{1/2}} = 69.3 \; \frac{m^{1/2}}{s} \, .$$

Example 8.4

A symmetrical notch has a section as shown. Prove that:
$$\dot{V} = C_D \sqrt{2g} \left[\tfrac{2}{3} BH^{3/2} + \tfrac{8}{15} (\tan \theta) H^{5/2} \right].$$

A notch of this form has all three sides 0.6 m long, the width at the top being 1.2 m. Find the discharge when $H = 0.52$ m.

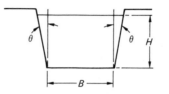

Answer

Reference to pages 112–113 will yield the given expressions. Note that the width is given as B here and as L in the notes and that $\theta/2$ on page 113 is equivalent to θ here. The trapezoidal channel (symmetrical about a vertical centre-line) is made up of a vee-notch and a rectangular notch and the discharge for each is additive to produce the above result.

By substitution, assuming C_D is 0.6,

$$\dot{V} = 0.6 \sqrt{2 \times 9.81} \ \frac{m^{1/2}}{s} \left[\left(\frac{2}{3} \times 0.6 \times 0.52^{3/2} \right) + \left(\frac{8}{15} \tan 30° \right) \times 0.52^{5/2} \right] m^{5/2}$$

$$= 2.66 \ (0.15 + 0.0602)$$

$$= 0.559 \ \frac{m^3}{s}.$$

Example 8.5

Water flows along a channel 1.5 m wide at rates varying from 0.006 m³/s to 0.7 m³/s. If the maximum allowable head is to be 125 mm, determine the depth and top width of a vee-notch capable of measuring the maximum flow if the co-efficient of discharge is 0.59.

If the apex of the notch is 1 m above the channel bed, what will be the effect of allowing for the channel approach velocity at the maximum flow condition?

Answer

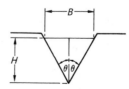

For a vee-notch (see page 113),

$$\dot{V} = C_D \sqrt{2g} \cdot \frac{8}{15} \tan \theta \cdot H^{5/2}.$$

Thus

$$\theta = \arctan \frac{15\dot{V}}{8\sqrt{2g}\,H^{5/2}\,C_{\mathrm{D}}} = \arctan \frac{15 \times 0.006 \; \frac{\mathrm{m}^3}{\mathrm{s}}}{8\sqrt{2 \times 9.81} \; \frac{\mathrm{m}^{1/2}}{\mathrm{s}} \times 0.125^{5/2} \; \mathrm{m}^{3/2} \times 0.59}$$

$$= \arctan 0.779 \; (37.9^\circ).$$

Thus for $\dot{V} = 0.7 \; \frac{\mathrm{m}^3}{\mathrm{s}}$,

$$H^{5/2} = \frac{15\dot{V}}{8\sqrt{2g} \cdot \tan\theta \cdot C_{\mathrm{D}}} = \frac{15 \times 0.7 \; \frac{\mathrm{m}^3}{\mathrm{s}}}{8\sqrt{2 \times 9.81} \; \frac{\mathrm{m}^{1/2}}{\mathrm{s}} \times 0.779 \times 0.59} = 0.6447 \; \mathrm{m}^{5/2},$$

$$H = 0.6447^{2/5} = 0.839 \; \mathrm{m},$$

$$B = 2H \tan\theta = 2 \times 0.839 \times 0.779 \; \mathrm{m} = 1.307 \; \mathrm{m}.$$

$$u_{\mathrm{approach}} = \frac{\dot{V}}{A} = \frac{0.7 \; \frac{\mathrm{m}^3}{\mathrm{s}}}{1.5 \; \mathrm{m} \times 1.84 \; \mathrm{m}} = 0.254 \; \frac{\mathrm{m}}{\mathrm{s}},$$

$$\frac{u_{\mathrm{app}}^2}{2g} = \frac{0.254^2}{2 \times 9.81} \; \mathrm{m} = 0.003\,28 \; \mathrm{m},$$

$$\dot{V} = \frac{2}{3} B \sqrt{2g} \left[\left(H + \frac{u_{\mathrm{a}}^2}{2g} \right)^{3/2} - \left(\frac{u_{\mathrm{a}}^2}{2g} \right)^{3/2} \right],$$

where u_{a} here (approach velocity) $= u_1$ on page 114,

$$\frac{\mathrm{d}\dot{V}}{\dot{V}} = \frac{5}{2}\left(\frac{u_{\mathrm{a}}^2/2g}{H + u_{\mathrm{a}}^2/2g} \right) = 2.5 \times \left(\frac{0.003\,28}{0.84 + 0.003\,28} \right) \times 100 \text{ per cent} = 0.97 \text{ per cent}.$$

Example 8.6

Water flows from a pond over a weir 3 m long to a depth of 0.25 m (two end contractions). It then flows along a level rectangular channel 2.5 m wide and over a second weir the width of the channel (no end contractions) whose crest is 0.3 m above the channel bed. Using the Francis formula for both weirs, calculate the depth of water over the 2.5 m weir.

Answer

Francis formula is equation 3 on page 114.

$$\dot{V} = 1.84\,(B - 0.1\,nH) \times H^{3/2} \qquad (n = \text{number of end contractions})$$

$$= 1.84\,(3 - 0.1 \times 2 \times 0.25) \times 0.25^{3/2} = 0.679 \; \frac{\mathrm{m}^3}{\mathrm{s}}.$$

Now $\dot{V} = 1.84 B H^{3/2}$,

or $$H = \left(\frac{\dot{V}}{1.84\,B} \right)^{2/3} = \left[\frac{0.679 \; \mathrm{m}^3/\mathrm{s}}{1.84 \; \frac{\mathrm{m}^{1/2}}{\mathrm{s}} \times 2.5 \; \mathrm{m}} \right]^{2/3} = 0.279 \; \mathrm{m}.$$

To correct for approach velocity u_a,

$$u_a = \frac{\dot{V}}{A} = \frac{0.679 \text{ m}^3/\text{s}}{2.5 \text{ m} \times (0.3 + 0.279) \text{ m}} = 0.469 \text{ m/s},$$

$$\frac{u_a^2}{2g} = \frac{0.469^2}{2 \times 9.81} = 0.0112 \text{ m}.$$

Thus $\quad 0.679 \frac{\text{m}^3}{\text{s}} = 1.84 \frac{\text{m}^{1/2}}{\text{s}} \times 2.5 \text{ m} \, [(H + 0.0112)^{3/2} - (0.0112)^{3/2}] \text{ m}^{3/2}$

or $\quad (H + 0.0112)^{3/2} = 0.147 + 0.001\,18 = 0.148 \text{ m}^{3/2},$

$$H + 0.0112 = 0.28 \text{ m},$$

$$H = 0.27 \text{ m}.$$

Example 8.7

In measuring the head of water flowing in a channel an error of 0.001 m is possible. Find the percentage error in an estimated discharge of 0.012 m³/s when using:
(a) a 90° triangular notch for which $\dot{V} = 1.41 H^{5/2}$,
(b) a rectangular notch 0.6 m long for which $\dot{V} = 1.84 L H^{3/2}$.
    Above what flow rate will this percentage error become greater for the triangular notch than for the rectangular one?

Answer

(a) $\quad H = \left(\frac{\dot{V}}{1.41}\right)^{2/5} = \left(\frac{0.012}{1.41}\right)^{0.4} = 0.149 \text{ m}.$

$$\frac{\text{d}\dot{V}}{\dot{V}} = \frac{\text{d}H}{H} \times 2.5 = \frac{0.001}{0.149} \times 2.5 \times 100 \text{ per cent} = 1.68 \text{ per cent}.$$

(b) $\quad H = \left(\frac{\dot{V}}{1.84 L}\right)^{2/3} = \left(\frac{0.012}{1.84 \times 0.6}\right)^{2/3} = 0.049 \text{ m}.$

$$\frac{\text{d}\dot{V}}{\dot{V}} = 1.5 \times \frac{0.001}{0.049} \times 100 \text{ per cent} = 3.06 \text{ per cent}.$$

(c) If

$$\left(\frac{\text{d}\dot{V}}{\dot{V}}\right)_{\text{vee}} = \left(\frac{\text{d}\dot{V}}{\dot{V}}\right)_{\text{rect}},$$

$$2.5 \frac{\text{d}H}{H_{\text{vee}}} = 1.5 \frac{\text{d}H}{H_{\text{rect}}},$$

i.e.

$$2.5 \left(\frac{\dot{V}}{1.84 \times 0.6}\right)^{2/3} = 1.5 \left(\frac{\dot{V}}{1.41}\right)^{2/5}.$$

$$\dot{V}^{4/15} = \frac{1.5}{2.5} \times \frac{1.104^{2/3}}{1.41^{2/5}} = 0.5586,$$

$$\dot{V} = 0.113 \frac{\text{m}^3}{\text{s}}.$$

Example 8.8

Explain the difference between an orifice and a notch.

A notch in an open channel is triangular (see diagram). Working from first principles, show that the volumetric flow rate of water through the notch is given by

$$\dot{V} = \tfrac{4}{15} C_\mathrm{D} \, \tan \theta \, \sqrt{2g} \, H^{3/2} \, (5L - 2H)$$

where H, the head over the notch, is always less than L.

Answer

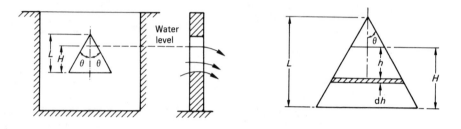

At depth h from Bernoulli's equation,

$$\text{ideal } u = \sqrt{2gh}.$$

Ideal $d\dot{V}$ through the shaded section = $u \times$ area.

Ideal $d\dot{V} = \sqrt{2gh} \cdot 2 \, [L - H + h] \, \tan \theta \, dh$,

$$\text{ideal } \dot{V} = \int_0^H 2\sqrt{2g} \, [(L-H)h^{1/2} + h^{3/2}] \, \tan \theta \, dh$$

$$= [(L-H)H^{3/2} \times \tfrac{2}{3} + \tfrac{2}{5} H^{5/2}] \; 2\sqrt{2g} \, \tan \theta$$

$$= H^{3/2} \left[\frac{10(L-H) + 6H}{15}\right] 2\sqrt{2g} \, \tan \theta$$

$$= H^{3/2} \left[\frac{10L - 4H}{15}\right] 2\sqrt{2g} \, \tan \theta = H^{3/2} \, [5L - 2H] \, \frac{4}{15}\sqrt{2g} \, \tan \theta.$$

$$\text{Actual } \dot{V} = C_\mathrm{D} \, \dot{V}_\text{ideal} = \tfrac{4}{15} C_\mathrm{D} \, \tan \theta \, \sqrt{2g} \, H^{3/2} \, [5L - 2H].$$

Exercises

1 Calculate the discharge in $\mathrm{m^3/s}$ for the channels whose cross-sectional shapes are depicted in the diagrams on page 121. Take the Chezy constant as $60 \; \mathrm{m^{1/2}}$ in each case and the slope as 1 in 3000. (*Answer:* $103.8 \; \mathrm{m^3/s}, 36.15 \; \mathrm{m^3/s}$)

2 Prove from first principles that the quantity $\dot{V}$ of a fluid flowing through a vee-notch is related to the notch angle θ and the head of fluid H at the notch by the following expression:

$$\dot{V} = \frac{8}{15} \, \tan \frac{\theta}{2} \, \sqrt{2g} \, H^{5/2}.$$

What modification would you make to this expression to allow for both the contraction of the vein and any energy degradation?

3 A vee-notch has a total angle of 60° and a coefficient of discharge of 0.60. Working from first principles, develop an expression for the discharge over the notch of the form:

$$\dot{V} = KH^n,$$

where H is the head in mm.

If an error of 2 mm is made in reading this head, what will be the percentage error in the calculated value of $\dot{V}$?

(*Answer:* $\dot{V} = 25.88 \times 10^{-6} H^{5/2}$, 0.4 per cent)

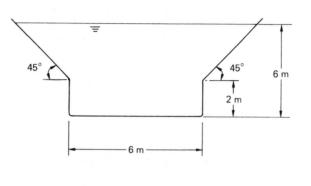

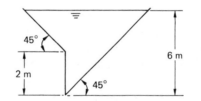

9 Steady Laminar Incompressible Viscous Flow

9.1 Incompressible Flow of Fluid Between Parallel Plates

Assumptions:

(a) Newtonian fluid (see Chapter 1).
(b) Constant absolute viscosity of fluid.
(c) No slip between fluid and plate surface.
(d) No pressure gradient across the fluid layer.
(e) Viscous forces much greater than inertia forces (i.e. no appreciable acceleration in the direction of flow).
(f) Gravitational or body forces negligible.

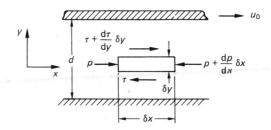

Consider a fluid layer of thickness δy and of length δx moving with velocity u and subject to the pressures shown in the diagram. Let the element have a depth b into the page.

The velocity gradient in the fluid gives rise to shear stresses as shown.

Nett force due to pressure on the element is given by

$$\left[p - \left\{p + \frac{\mathrm{d}p}{\mathrm{d}x}\,\delta x\right\}\right] b\,\delta y = -b\,\frac{\mathrm{d}p}{\mathrm{d}x}\,\delta x\,\delta y.$$

Resistance due to viscous shear
$$= \left[\tau - \left\{\tau + \frac{\mathrm{d}\tau}{\mathrm{d}y}\,\delta y\right\}\right] b\,\delta x = -b\,\frac{\mathrm{d}\tau}{\mathrm{d}y}\,\delta x\,\delta y.$$

For assumed fully developed flow these two are equal, giving:

$$\frac{\mathrm{d}p}{\mathrm{d}x} = \frac{\mathrm{d}\tau}{\mathrm{d}y}.$$

For a Newtonian fluid, $\tau = \mu \dfrac{\mathrm{d}u}{\mathrm{d}y}$ (page 3).

or $\dfrac{\mathrm{d}p}{\mathrm{d}x} = \mu \dfrac{\mathrm{d}^2 u}{\mathrm{d}y^2}.$ ... (1)

9.2 Shear Stress in a Circular Pipe with Steady Flow

The effect of viscosity is to lead to a degradation of mechanical energy and thus to a decrease in piezometric pressure $(p + \rho gz)$. This is directly related to the shear stress at the flow boundaries.

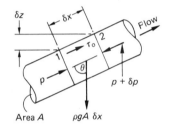

For a pipe running full, of radius r (suffix o refers to solid boundary),

net force on fluid element in flow direction = 0

(since velocity distribution over cross-section is constant relative to x in a fully established steady flow).

$$pA - (p + \delta p)\,A - \rho gA\,\delta x \cos \theta + \tau_o P\,\delta x = 0$$

(A = area, P = perimeter).

Now $\delta x \cos \theta = \delta z.$

Thus $\tau_o P\,\delta x = A\,(\delta p + \rho g\,\delta z) = A \times p^*$ $(p^* = p + \rho gz)$

and as $\delta x \to 0$, $\tau_o = \dfrac{A}{P}\dfrac{\mathrm{d}p^*}{\mathrm{d}x}.$

Now decrease in $p^* = -\delta p^* = \rho g\,\delta h_f$ (h_f = energy degradations).

Thus $\dfrac{\mathrm{d}h_f}{\mathrm{d}x} = -\dfrac{1}{\rho g}\dfrac{\mathrm{d}p^*}{\mathrm{d}x} = -\dfrac{\tau_o P}{\rho gA}.$

Darcy experimented and found that $f = -\dfrac{2g}{u_m^2}\dfrac{\tau_o P}{\rho gA}$,

or $f = \dfrac{\tau_o}{\frac{1}{2}\rho u_m^2}.$ ... (2)

9.3 Laminar Flow Through a Circular Pipe

Refer to the diagram, in which fluid flows in viscous (laminar) flow along a pipe of radius R. The assumptions made are similar to those listed previously for parallel plates.

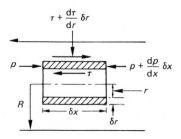

We can consider both the cylindrical element of radius r and the annular element of thickness δr. The net force on the element is zero.

Cylindrical element

The pressure and viscous forces are in balance:

$$\left[p + \frac{dp}{dx}\, \delta x - p \right] \cdot \pi r^2 = \tau 2\pi r\, \delta x$$

or
$$r\, \frac{dp}{dx} = 2\tau. \qquad\qquad \ldots (3)$$

Annular element

$$\left[p + \frac{dp}{dx}\, \delta x - p \right] 2\pi r\, \delta r = \left[\left(\tau + \frac{d\tau}{dr}\, \delta r \right) 2\pi (r + \delta r) - \tau 2\pi r \right] \delta x$$

or
$$r\, \frac{dp}{dx} = \tau + r\, \frac{d\tau}{dr}.$$

Integrating with respect to r gives

$$\left(\frac{dp}{dx} \right) \left(\frac{r^2}{2} \right) = \tau r + c \qquad \text{(and } c = 0 \text{ when } r = 0).$$

Thus
$$r\, \frac{dp}{dx} = 2\tau. \qquad\qquad \ldots (4)$$

Newton's law of viscosity

$$\tau = \mu\, \frac{du}{dr} = \frac{r}{2}\, \frac{dp}{dx},$$

and integrating with respect to r gives

$$\mu u = \frac{r^2}{4}\, \frac{dp}{dx} + D. \qquad \text{(but } r = R \text{ if } u = 0).$$

Thus
$$D = -\frac{R^2}{4}\frac{\mathrm{d}p}{\mathrm{d}x},$$

$$u = \frac{1}{4\mu}\frac{\mathrm{d}p}{\mathrm{d}x}(r^2 - R^2), \qquad \ldots (5)$$

$\left(\dfrac{\mathrm{d}p}{\mathrm{d}x}\text{ is negative in flow direction}\right)$.

i.e.
$$u = \frac{\Delta p}{4\mu L}(R^2 - r^2), \qquad \ldots (6)$$

a parabolic distribution.

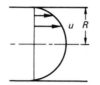

$$\dot{V} = \int_0^R 2\pi r u\,\mathrm{d}r = \int_0^R \frac{2\pi\,\Delta p\,\rho r}{4\mu L}(R^2 - r^2)\,\mathrm{d}r = \frac{\pi\Delta\rho}{2\mu L}\left(\frac{R^2 r^2}{2} - \frac{r^4}{4}\right)_0^R$$

$$= \frac{\pi\,\Delta p\,R^4}{8\mu L} = \frac{\pi\,\Delta p\,D^4}{128\mu L} \qquad \ldots (7)$$

(Hagan–Poiseuille relation).

$$u_\mathrm{m} = \frac{\dot{V}}{A} = \frac{\pi\,\Delta p\,R^4}{\pi R^2\,8\mu L} = \frac{\Delta p\,R^2}{8\mu L} = \frac{D^2\,\Delta p}{32\mu L}. \qquad \ldots (8)$$

$\hat{u}$ at $\dfrac{\mathrm{d}u}{\mathrm{d}r} = 0$ (at $r = 0$) is given by

$$\hat{u} = \frac{\Delta p\,R^2}{4\mu L} = 2u_\mathrm{m} \qquad \ldots (8a)$$

or
$$u = \hat{u}\left(1 - \frac{r^2}{R^2}\right). \qquad \ldots (9)$$

Furthermore,
$$f = \frac{\tau_0}{\frac{1}{2}\rho u_\mathrm{m}^2}$$

and at the boundary
$$\tau_\mathrm{o} = \mu\left(\frac{\mathrm{d}u}{\mathrm{d}r}\right)_R = \frac{R\,\Delta p}{2L} \qquad \text{(for a pipe of radius } R\text{)};$$

$$f = \frac{\dfrac{R\,\Delta p}{L}}{\frac{1}{2}\rho u_\mathrm{m}^2} = \frac{R}{2}\frac{8\mu u_\mathrm{m}}{R^2}\frac{2}{\rho u_\mathrm{m}^2} = \frac{8\mu}{\rho u_\mathrm{m}R} = \frac{16\mu}{\rho u_\mathrm{m}D}$$

$$= \frac{16}{(Re)}. \qquad \ldots (10)$$

9.4 Laminar Flow Between Stationary Parallel Pipes

(a) Constant Pressure Along the Channel

i.e.
$$\frac{dp}{dx} = 0.$$

In equation 1,
$$\frac{d^2 u}{dy^2} = 0$$

and integrating twice with respect to y gives

$$u = Cy + D \qquad \text{(C and D constants)}$$

Boundary conditions:
$$u = 0; \qquad y = 0;$$
$$u = u_o; \qquad y = d.$$

substitution gives
$$u = \frac{u_o y}{d}. \qquad \qquad \ldots (11)$$

Thus (see diagram) u is a linear function of y.

Now
$$\frac{du}{dy} = \frac{u_o}{d}$$

and
$$\tau = \mu \frac{du}{dy} = \mu \frac{u_0}{d} = \text{a constant.}$$

(b) Constant Pressure Gradient Between Two Stationary Plates

i.e. $\dfrac{dp}{dx} = \text{constant};$ $u_0 = 0.$

Integrating equation 1 twice with respect to y, we get

$$\frac{1}{2} \frac{dp}{dx} y^2 + Cy + D = \mu u.$$

Boundary conditions: $u = 0;$ $y = 0.$ Thus $D = 0.$

$$u = 0; \qquad y = d; \qquad \text{and } C = -\frac{1}{2} \frac{dp}{dx} d.$$

Thus $u = \dfrac{1}{2\mu} \dfrac{dp}{dx} (y^2 - yd).$ $\qquad \qquad \ldots (12)$

Maximum velocity $\hat{u}$ exists when $\dfrac{du}{dy} = 0$

or when

$$\frac{1}{2\mu} \frac{dp}{dx} (2y - d) = 0,$$

$$y = \frac{d}{2},$$

and

$$\hat{u} = - \frac{d^2}{8\mu} \frac{dp}{dx}, \qquad \text{a parabolic profile.}$$

Volume flow rate between plates is

$$\dot{V} = b \int_0^d u\, dy = b \int_0^d \frac{1}{2\mu} \frac{dp}{dx} (y^2 - yd)\, dy$$

$$= \frac{b}{2\mu} \frac{dp}{dx} \left[\frac{y^3}{3} - \frac{y^2 d}{2} \right]_0^d = - \frac{bd^3}{12\mu} \frac{dp}{dx}. \qquad \ldots (13)$$

Mean velocity

$$\left. \begin{array}{c} u_m = \dfrac{\dot{V}}{bd} = \dfrac{-d^2}{12\mu} \dfrac{dp}{dx}, \\[2mm] \hat{u} = \frac{3}{2} u_m. \end{array} \right\} \qquad \ldots (14)$$

Note that dp/dx is negative in the flow direction and if we replace dp/dx by $\Delta p/L$, where Δp is the pressure drop over length L,

$$u_m = \frac{d^2 \Delta p}{12\mu L}. \qquad \ldots (14a)$$

(c) Constant Pressure Gradient, One Surface Moving at Constant Velocity u_0

$$\frac{dp}{dx} = \text{constant}; \quad u_o = \text{constant}.$$

Integrating twice with respect to y gives

$$\frac{1}{2} \frac{dp}{dx} y^2 + Cy + D = \mu u.$$

Boundary conditions: $u = 0$; $y = 0$; $D = 0$.

$u = u_o$; $y = d$; $C = \dfrac{\mu u_o}{d} = - \dfrac{1}{2} \dfrac{dp}{dx} d$.

$$u = \frac{u_o y}{d} + \frac{1}{2\mu} \frac{dp}{dx} (y^2 - yd). \qquad \ldots (15)$$

127

Combination of Cases (a) and (b)

For $\dfrac{dp}{dx} = 0$ we get a linear velocity profile,

for $\dfrac{dp}{dx}$ = a positive value,

$$\frac{du}{dy} = \frac{u_0}{d} - \frac{1}{2\mu}\frac{dp}{dx}(2y - d) = 0 \text{ for a stationary value,}$$

i.e. $y = \dfrac{d}{2} - \dfrac{\mu u_0}{\dfrac{dp}{dx}\,d}$

and $\dfrac{d^2 u}{dy^2} = \dfrac{1}{\mu}\dfrac{dp}{dx}$ (a minimum value).

Thus there is a reverse flow in the vicinity of the stationary wall.

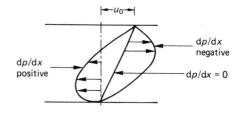

$\dot{V}$ between the plates $= b\displaystyle\int_0^d u\,dy$;

$$\dot{V} = b\int_0^d \frac{u_0 y}{d} + \frac{1}{2\mu}\frac{dp}{dx}(y^2 - yd)\,dy$$

$$= b\left[\frac{u_0 d}{2} - \frac{d^3\dfrac{dp}{dx}}{12\mu}\right]. \qquad\qquad \dots(16)$$

$$u_\mathrm{m} = \frac{\dot{V}}{bd} = \frac{u_0}{2} - \frac{d^2\dfrac{dp}{dx}}{12\mu}, \qquad\qquad \dots(17)$$

$$\tau = \mu\frac{du}{dy} = \frac{\mu u_0}{d} + \frac{1}{2}\frac{dp}{dx}(2y - d).$$

And at moving surface $y = d$,

thus

$$\tau = \frac{\mu u_0}{d} + \frac{d}{2}\frac{dp}{dx}. \qquad\qquad \dots(18)$$

Now

$$f = \frac{\tau_0}{\frac{1}{2}\rho u_\mathrm{m}^2},$$

and if

$$\frac{dp}{dx} = \frac{-\Delta p}{L},$$

where Δp is pressure drop over length L,

$$u = \frac{\Delta p}{2\mu L}(y^2 - yd)$$

(equation 12) and for $y = 0$ or d

$$\frac{du}{dy} = \pm \frac{d\,\Delta p}{2\,\mu L}.$$

Thus

$$\tau_o = \mu \frac{du}{dy} = \pm \frac{d\,\Delta p}{2L}.$$

Taking positive values of τ_o and using $\mu_m = \frac{d^2\,\Delta p}{12\mu L}$ (equation 14a), we get

$$\tau_o = \frac{d}{2}\left(\frac{12\mu u_m}{d}\right) = \frac{6\mu u_m}{d},$$

$$f = \frac{\frac{6\mu u_m}{d}}{\frac{1}{2}\rho u_m^2} = \frac{12\mu}{\rho u_m d} = \frac{12}{(Re)}. \qquad \ldots (19)$$

9.5 Dashpot

In this device a movement of a piston forces oil through a clearance gap between the piston and cylinder. As the clearance is very much less than the piston diameter the problem approximates to the parallel-plate case previously treated. The flow is assumed steady, and fully developed.

Here dp/dx is positive, and is usually large compared with u_P.

Let the cylinder pressure be p and the piston length L.

$$\frac{dp}{dx} = \frac{p}{L},$$

and, referring to equation 16,

$$\dot{V} = b\left[\frac{u_P t}{2} - \frac{t^2 p}{12\mu L}\right] \qquad \text{and is negative}$$

where $b = \pi D$, t replaces d in 16,

i.e.

$$\dot{V} = \pi D\left[\frac{u_P t}{2} - \frac{t^3 p}{12\mu L}\right].$$

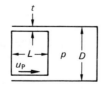

However the discharge $\dot{V}$ = piston displacement rate = $-\frac{\pi D^2}{4} u_P$,

i.e.

$$\pi D\left[\frac{u_P t}{2} - \frac{t^3 p}{12\mu L}\right] = -\frac{\pi D^2}{4} u_P$$

from which
$$u_P = \frac{\dfrac{t^3 p}{12\mu L}}{\dfrac{t}{2} + \dfrac{D}{4}} = \frac{t^3 p}{3\mu L\,(2t + D)}.\qquad \ldots (20)$$

Since the piston speed u_P is small compared with $\dfrac{dp}{dx}$, the simplified flow rate is

$$\dot{V} = - \frac{\pi D t^3 p}{12\mu L},$$

giving
$$u_P = \frac{\dot{V}4}{\pi D^2} = \frac{t^3 p}{3\mu L D}$$

and
$$\frac{p}{L} = \frac{3\mu\,(2t + D)\,u_P}{t^3} = \frac{3\mu D u_P}{t^3}\qquad (t \ll D).$$

Also, the power to move the piston = force × velocity.

$$\therefore \text{power} = p\,\frac{\pi D^2}{4}\,u_P = -\dot{V}p \qquad \text{(which is input power)}. \qquad \ldots (21)$$

9.6 Worked Examples

Example 9.1

A plane flat plate, 600 mm long and 600 mm wide, slides down an inclined plane with a constant velocity of 180 mm/s as shown in the diagram.

Determine the viscosity of the oil in the oil film between the plate and the inclined plane. The plate has a mass of 11 kg.

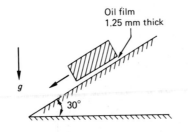
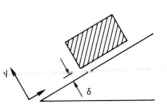

Answer

In the oil film the flow is steady, with no pressure gradient.

$$\text{Velocity gradient} = \frac{du}{dy} = \frac{0.18\ \dfrac{m}{s}}{0.001\,25\ m} = 144\ \frac{1}{s} \qquad (dy = \delta = 0.001\,25\ m).$$

Shear force at plate surface $F = \mu \left(\dfrac{du}{dy}\right)_\delta \times \text{area}$

$$= \mu \times 144\ \frac{1}{s} \times 0.6^2\ m^2 = 51.84\ \mu\ \frac{m^2}{s}.$$

But if mg is weight of plate acting down through its centroid then

$$F = mg \sin \theta = 11 \text{ kg} \times 9.81 \ \frac{\text{m}}{\text{s}} \ \sin 30° \left[\frac{\text{N s}^2}{\text{kg m}} \right] = 53.955 \text{ N}.$$

Thus $\quad \mu = \dfrac{53.955 \text{ N}}{51.84 \ \dfrac{\text{m}^2}{\text{s}}} = 1.0408 \ \dfrac{\text{N s}}{\text{m}^2}$.

Example 9.2

A cylinder is of diameter 0.2 m and is 1 m long. It is concentric with a pipe of inside diameter 0.206 m.

   If, between the cylinder and the pipe, there is an oil film, what force is required to move the cylinder along the pipe at a velocity of 1 m/s? Take the kinematic viscosity of the oil as 6×10^{-4} m^2/s and its specific gravity as 0.92 and assume steady flow.

Answer

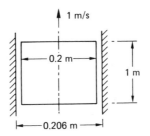

Film thickness = 0.003 m,

$$\frac{du}{dy} = \frac{1 \ \dfrac{\text{m}}{\text{s}}}{0.003 \text{ m}} = 33.3 \ \frac{1}{\text{s}}.$$

Shear stress at cylinder surface $\tau = \mu \ \dfrac{du}{dy}$.

Shear force = $\tau \times$ area $= \tau \times 2\pi rL$,

and neglecting weight of cylinder,

$$\text{shear force} = \mu \ \frac{du}{dy} \ \pi \, dL = \rho v \ \frac{du}{dy} \ \pi \, dL$$

$$= 920 \ \frac{\text{kg}}{\text{m}^3} \times 6 \times 10^{-4} \ \frac{\text{m}^2}{\text{s}} \times 333.3 \ \frac{1}{\text{s}} \times \pi \times 0.2 \text{ m} \times 1 \text{ m} \left[\frac{\text{N s}^2}{\text{kg m}} \right]$$

$$= 115.6 \text{ N}.$$

Example 9.3

Show that for viscous flow in one direction between two plane and parallel surfaces separated by a distance t, the loss of pressure per unit length in the direction of flow is given by

$$\frac{p}{L} = \frac{12\mu u_m}{t^2},$$

where μ = the coefficient of absolute viscosity,

$\quad\quad\; \mu_m$ = the mean velocity of flow.

A mass of 20 tonne is lifted by a piston of diameter 399 mm and length 150 mm in a cylinder of diameter 400 mm by pumping in oil of absolute viscosity 2 N s/m² below the piston.

Calculate the leakage past the piston in m³/h and the power required to keep it at a constant level when the axis of the cylinder is vertical. Assume the piston mass is negligible.

Answer

The required expression is equation 13 recast.

$$p = \frac{W}{A} = \frac{mg}{A} = \frac{20 \times 10^3 \text{ kg} \times 9.81 \frac{m}{s^2} \times 4}{\pi \times 0.399^2 \text{ m}^2} \left[\frac{\text{N s}^2}{\text{kg m}}\right] = 15.691 \times 10^5 \frac{\text{N}}{\text{m}^2}.$$

$$\dot V = \frac{-\pi D t^3 p}{12\mu L} \quad\quad \left(\text{equation 13, where } b = \pi d; t = d; \frac{dp}{dx} = \frac{p}{L}\right)$$

$$= \frac{-\pi \times 0.399 \text{ m} \times 0.5^3 \text{ m}^3 \times 10^{-9} \times 15.691 \times 10^5 \frac{\text{N}}{\text{m}^2}}{12 \times 2 \frac{\text{N s}}{\text{m}^2} \times 0.15 \text{ m}} \left[\frac{3600 \text{ s}}{\text{h}}\right] = 0.246 \frac{\text{m}^3}{\text{h}}.$$

$$\dot W = -\dot V p = 0.246 \frac{\text{m}^3}{\text{n}} \times 15.691 \frac{\text{N}}{\text{m}^2} \times 10^5 \left[\frac{\text{h}}{3600 \text{ s}}\right] \left[\frac{\text{J}}{\text{N m}}\right] = 107.22 \text{ W}.$$

Example 9.4

Show that in the laminar flow of an incompressible fluid through a tube of radius a, the axial velocity at radius r is given by

$$u_o \left[1 - \frac{r^2}{a^2}\right],$$

where u_o is the velocity at the axis of the tube. Show that the kinetic energy per unit mass flow rate is given by u_m^2 where u_m is the mean velocity of flow.

Answer

Equation 6: $\quad\quad\quad u = \frac{1}{4\mu} \frac{dp}{dx} (r^2 - R^2),\quad\quad$ and putting $r = r, R = a$,

$$u = -\frac{dp}{dx} \frac{1}{4\mu} (a^2 - r^2);$$

at tube axis ($r = 0$), $\quad u = \hat u = u_o$ as given here,

and $\quad\quad\quad\quad\quad\quad u = \hat u \left(1 - \frac{r^2}{a^2}\right) = u_o \left(1 - \frac{r^2}{a^2}\right).$

Kinetic energy of annulus of fluid $= \frac{1}{2} mu^2$

$$= \frac{1}{2} \rho 2\pi r \, dr \, u \cdot u^2.$$

And total kinetic energy $= \int_0^a \rho \pi r \, (2u_m)^3 \left[1 - \frac{r^2}{a^2} \right]^3 dr$

$$= 8\pi\rho \, u_m \int_0^a \left[r - \frac{3r^3}{a} + \frac{3r^5}{a} - \frac{r^7}{a} \right] dr$$

$$= \pi\rho a^2 \, u_m^3 \, .$$

Total k.e./unit mass flow rate $= \dfrac{\pi\rho a^2 \, u_m^3}{\pi\rho a^2 \, u_m} = u_m^2 \, .$

Example 9.5

A plunger of diameter 200 mm and of length 200 mm slides in a vertical cylinder with a radial clearance of 0.25 mm and carries a load of 500 kN. If the cylinder is filled with oil of absolute viscosity 0.15 N s/m^2, estimate the leakage past the piston and the power necessary to maintain the load. At what rate will the plunger sink when the oil supply valve is shut?

Assume that $\qquad u_m = \dfrac{u_P}{2} - \dfrac{t^2}{12\mu} \left(\dfrac{dp}{dx} \right) \, ,$

where u_P = plunger velocity.

Answer

$$\Delta p = \frac{500\,000 \text{ N}}{\pi \times 0.1^2 \text{ m}^2} = 15.92 \times 10^6 \ \frac{\text{N}}{\text{m}^2} \, ,$$

$$\Delta x = 0.2 \text{ m},$$

$$\frac{\Delta p}{\Delta x} = \frac{15.92}{0.2} \times 10^6 \ \frac{\text{N}}{\text{m}^3} = 79.55 \times 10^6 \ \frac{\text{N}}{\text{m}^3} \, ,$$

when $u_P = 0$, $u_m = \dfrac{t^2}{12\mu} \dfrac{dp}{dx} = \dfrac{0.25 \text{ m}^2 \times \text{m}^2}{10^6 \times 12 \times 0.15 \text{ N s}} \times 79.55 \times 10^6 \ \dfrac{\text{N}}{\text{m}^3}$

or $\quad u_m = 2.762 \ \dfrac{\text{m}}{\text{s}} \, ,$

$$\dot{V} = u_m \, \pi Dt = 2.762 \ \frac{\text{m}}{\text{s}} \times \pi \times 0.2 \text{ m} \times \frac{0.25 \text{ m}}{10^3} = 4.339 \times 10^{-4} \ \frac{\text{m}^3}{\text{s}} \quad (\text{OUT}).$$

Power $= -\dot{V}p = 4.339 \times 10^{-4} \ \dfrac{\text{m}^3}{\text{s}} \times 15.92 \times 10^6 \dfrac{\text{N}}{\text{m}^2} \left[\dfrac{\text{kW s}}{10^3 \text{ N m}} \right]$

$$= 6.91 \text{ kW}.$$

Now $\quad = u_P \times \frac{1}{4} \, (\pi D)^2 = u_m \, \pi Dt,$

or $\quad u_P = u_m \dfrac{4t}{D} = 2.762 \ \dfrac{\text{m}}{\text{s}} \times \dfrac{4 \times 0.25 \text{ m}}{0.2 \text{ m} \times 10^3} = 0.0138 \ \dfrac{\text{m}}{\text{s}} \, .$

Example 9.6

Show that in the laminar flow of a fluid of absolute viscosity μ through a circular pipe of length L and diameter d, the quantity flowing is given by

$$\dot{V} = \frac{\pi p d^4}{128\mu L} \, ,$$

where p is the pressure difference over length L.

Oil of specific gravity 0.9 and absolute viscosity 0.128 N s/m² is pumped through a pipe of diameter 75 mm at the rate of 40 m³/h. Calculate the Reynolds number for the flow and determine the power necessary to pump the oil through a length of 300 m of pipe which rises vertically by 15 m. Ignore entry and exit velocities and energy degradation at these stations.

Answer

Equation 7 refers to first part (Hagan–Poiseuille) (putting $p = \Delta p$).

$$u = \frac{\dot{V}}{A} = 40 \ \frac{m^3}{h} \left[\frac{h}{3600 \ s} \right] \frac{4}{\pi \times 0.075^2 \ m^2} = 2.515 \ \frac{m}{s}.$$

$$(Re) = \frac{\rho u d}{\mu} = \frac{900 \ \frac{kg}{m^3} \times 2.515 \ \frac{m}{s} \times 0.075 \ m \ m^2}{0.128 \ N \ s} \left[\frac{N \ s^2}{kg \ m} \right] = 1326.$$

$$p = \frac{128 \dot{V} \mu L}{\pi D^4} = \frac{128 \times 40 \ \frac{m^3}{h} \left[\frac{h}{3600 \ s} \right] \times 0.0128 \ \frac{N \ s}{m^2} \times 300 \ m}{\pi \times 0.075^4 \ m^4}$$

$$= 5.5 \times 10^5 \ \frac{N}{m^2}.$$

$$\text{Equivalent head} = \frac{p}{\rho g} = \frac{5.5 \times 10^5 \ \frac{N}{m^2}}{900 \ \frac{kg}{m^3} \times 9.81 \ \frac{m}{s^2}} \left[\frac{kg \ m}{N \ s^2} \right]$$

$$= 62.29 \ \text{m oil}.$$

$$\text{Total head } H = 62.29 + 15 = 77.29 \ \text{m oil}.$$

Power $= \dot{m}gh = \rho \dot{V} g H$.

$$= 40 \ \frac{m^3}{h} \left[\frac{h}{3600 \ s} \right] 900 \ \frac{kg}{m^3} \times 9.81 \ \frac{m}{s^2} \times 77.29 \ m \left[\frac{N \ s^2}{kg \ m} \right] \left[\frac{kW \ s}{10^3 \ N \ m} \right]$$

$$= 7.58 \ \text{kW}.$$

Example 9.7

If a circular disc of diameter d is rotated in a liquid of absolute viscosity μ at a small distance Δh from a fixed surface, derive an expression for the torque τ necessary to maintain an angular velocity ω. Neglect effects due to radial acceleration.

The thrust of a shaft is taken by a collar bearing provided with a forced lubrication system which maintains a film of oil of uniform thickness between the surface of the collar and the surface of the bearing; the external and internal diameters of the collar are 0.15 m and 0.1 m respectively. If the thickness of the oil film which separates the surfaces is 0.25 mm and the coefficient of viscosity is 0.091 N s/m², determine the power lost in overcoming friction when the shaft is running at 300 rev/min.

Answer

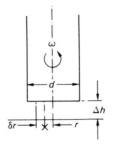

At radius r,
$$\frac{du}{dy} = \frac{\omega r}{\Delta h},$$

$$\tau r = \mu \frac{\omega r}{\Delta h} \qquad \text{(Newton).}$$

Force on element $\delta r = \delta F = \dfrac{\mu \omega r \, (2\pi r) \, \delta r}{\Delta h}.$

Torque on element $= \delta \tau = \delta F \times r = \dfrac{\mu \omega \, (2\pi r^3) \, \delta r}{\Delta h}$

Total torque $\tau = \displaystyle\int_{r=0}^{r=r} \delta \tau = \frac{2\pi \mu \omega}{\Delta h} \int_0^r r^3 \, dr$

$$= \frac{2\pi \mu \omega r^4}{\Delta h \cdot 4} = \frac{\pi}{32} \frac{\mu \omega d^4}{\Delta h}.$$

For a collar bearing, $\qquad \tau = \dfrac{\pi}{32} \dfrac{\mu \omega}{h} (d_2^4 - d_1^4),$

where
$$\omega = \left[\frac{2\pi \text{ radian}}{\text{rev}}\right] \left[\frac{300 \text{ rev}}{\text{min}}\right] \left[\frac{\text{min}}{60 \text{ s}}\right] = 10\pi \frac{1}{\text{s}};$$

$$\tau = \frac{\pi}{32} \times 0.091 \frac{\text{N s}}{\text{m}^2} \times 10\pi \frac{1}{\text{s}} \frac{10^4}{2.5 \text{ m}} [0.15^4 - 0.1^4] \text{ m}^4$$

$$= 0.456 \text{ N m}.$$

Power $= \tau \omega = 0.456 \text{ N m} \times 10\pi \dfrac{1}{\text{s}} \left[\dfrac{\text{W s}}{\text{N m}}\right] = 14.33 \text{ W.}$

Example 9.8

Calculate the oil flow rate, mean temperature rise and power loss in the oil film of a journal bearing, 100 mm long and of diameter 100 mm, in which the radial clearance is 0.10 mm when the shaft is concentric, the shaft speed being 400 radian/s. Assume the oil to have a constant kinematic viscosity of 25 centistokes, a specific gravity of 0.9 and a specific heat capacity of 1.88 kJ/kg K. Neglect leakage. State, but do not evaluate, what difference there is if the length/diameter ratio of the bearing is (a) 2/3 (b) 1/3. (*Note:* 1 centistokes (cSt) $= 10^{-6}$ m^2/s.)

Answer

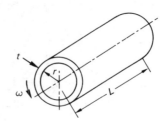

Mean velocity in oil film = $\dfrac{\omega r}{2}$.

Flow rate $\dot{V} = \dfrac{\omega r}{2} tL$ (for no leakage)

$$= \frac{400}{2} \frac{1}{s} \times \frac{100 \text{ mm}}{2} \times 0.1 \text{ mm} \times 100 \text{ mm} = 10^5 \frac{\text{mm}^3}{s}.$$

Heat transfer to the oil $= \dot{Q} = \dot{m}c\,\Delta T = \rho\,\dfrac{\omega r}{2}\,tL \times c\,\Delta T.$

Heat generated $= \dot{Q} =$ force × velocity $= \tau$ × area × velocity

$$= \mu\,\frac{du}{dy}\,\pi dL \cdot \omega r = \mu\,\frac{\omega r}{t}\,\pi dL \cdot \omega r$$

Thus $\rho\omega r\,tLc\,\Delta T = \mu\,\dfrac{\omega r}{2}\,\pi dL \cdot \omega r.$ $\Biggr\}$ Assuming no heat transfer to shaft or bearing.

Thus $\Delta T = \dfrac{\pi\mu\omega d^2}{\rho c t^2} = \dfrac{\pi\nu\omega d^2}{c t^2}$

$$= \frac{\pi \times 25 \times 10^{-6}\,\dfrac{\text{m}^2}{\text{s}} \times 400\,\dfrac{1}{\text{s}} \times 0.1^2\,\text{m}^2}{1.88\,\dfrac{\text{kJ}}{\text{kg K}} \times (0.1 \times 10^{-3})^2\,\text{m}^2} \left[\frac{\text{kJ}}{10^3\,\text{N m}}\right]\left[\frac{\text{N s}^2}{\text{kg m}}\right]$$

$= 16.7 \text{ K}.$

Power loss $= \dot{m}\,c\,\Delta T = \rho\dot{V}c\,\Delta T$

$$= 900\,\frac{\text{kg}}{\text{m}^3} \times \frac{10^5}{10^9}\,\frac{\text{m}^3}{\text{s}} \times 1.88\,\frac{\text{kJ}}{\text{kg K}} \times 16.7 \text{ K} \left[\frac{10^3\,\text{N m}}{\text{kJ}}\right]$$

$= 2\,825.6 \text{ W} \ (2.826 \text{ kW})$

For a given diameter, as L/D decreases heat transfer decreases. Thus with unaltered ΔT the power loss decreases.

Example 9.9

A uniform oil film is maintained over a chute 500 mm wide, which is inclined at 60° to the horizontal when the rate of flow is 0.2 m³/min. Assuming laminar flow and ignoring side effects and air resistance, derive
(a) an expression for the velocity distribution in the oil film,

(b) the ratio maximum velocity/mean velocity,
(c) the thickness of the oil film.

Take the kinematic viscosity of the oil as 0.5×10^{-4} m^2/s.

Answer

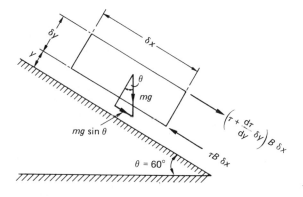

Let the width of the element be B (into page).
Let the mass of the element be m.
For fully developed flow (no acceleration; no net force on film),

$$mg \sin \theta + \left(\tau + \frac{d\tau}{dy}\, \delta y\right) B\, \delta x - \tau B\, \delta x = 0,$$

$$B\, \delta y\, \delta x\, \rho g \sin \theta + \left(\tau + \frac{d\tau}{dy}\, \delta y\right) B\, \delta x - \tau B\, \delta x = 0.$$

Thus
$$\rho g \sin \theta + \frac{d\tau}{dy} = 0$$

But $\quad \tau = \mu \dfrac{du}{dy} \quad$ and $\quad \dfrac{d\tau}{dy} = \mu \dfrac{d^2 u}{dy^2} = -\rho g \sin \theta.$

Integration gives $\quad \mu \dfrac{du}{dy} = \rho g \sin \theta\, y + \mathrm{A}.$

Boundary conditions: $\quad \dfrac{du}{dy} = 0$ at free surface $\qquad$ (since $\tau = 0$ here).

Thus $\qquad \mathrm{A} = \rho g \sin \theta\, t \qquad$ (t = film thickness).

Thus $\qquad \mu \dfrac{du}{dy} = -\rho g \sin \theta\, y + \rho g \sin \theta\, t = \rho g \sin \theta\, (t - y).$

Thus, integrating again, we get

$$\mu u = \rho g \sin \theta \left(ty - \frac{y^2}{2}\right) + \mathrm{C}.$$

Boundary conditions: $\qquad u = 0; \qquad y = 0; \qquad$ (solid surface)

Thus $\qquad \mathrm{C} = 0.$

Thus $\qquad \mu u = \rho g \sin \theta \left(ty - \dfrac{y^2}{2}\right)$

or
$$u = \frac{\rho g \sin \theta}{\mu} \left(ty - \frac{y^2}{2} \right). \qquad \ldots (a)$$

$\hat{u}$ at $y = t$.

Thus
$$\mu \hat{u} = \rho g \sin \theta \, \frac{t^2}{2},$$

$$\hat{u} = \frac{\rho g \sin \theta}{\mu} \, \frac{t^2}{2}.$$

$$\dot{V} = B \int_0^t u \, dy = B \int_0^t \frac{\rho g \sin \theta}{\mu} \left(ty - \frac{y^2}{2} \right) dy$$

$$= \frac{B \rho g \sin \theta}{\mu} \left[\frac{ty^2}{2} - \frac{y^3}{6} \right]_0^t = \frac{B \rho g \sin \theta}{\mu} \, \frac{t^3}{3}.$$

But
$$\dot{V} = Bt \, u_\mathrm{m}.$$

Thus
$$u_\mathrm{m} = \frac{\rho g \sin \theta}{\mu} \, \frac{t^2}{3},$$

$$\frac{\hat{u}}{u_\mathrm{m}} = \frac{t^2/2}{t^2/3} = \frac{3}{2}. \qquad \ldots (b)$$

$$\dot{V} = 0.2 \, \frac{\mathrm{m}^3}{\mathrm{mm}} \left[\frac{\mathrm{min}}{60 \, \mathrm{s}} \right] = \frac{1}{300} \, \frac{\mathrm{m}^3}{\mathrm{s}} = \frac{B \rho g \sin \theta}{\mu} \left(\frac{t^3}{3} \right).$$

Thus
$$t = \left[\frac{1}{300} \, \frac{\mathrm{m}^3}{\mathrm{s}} \times \frac{0.5}{10^4} \, \frac{\mathrm{m}^2}{\mathrm{s}} \times \frac{3}{0.5 \, \mathrm{m}} \, \frac{\mathrm{s}^2}{9.81 \, \mathrm{m}} \, \frac{1}{0.866} \right]^{1/3}$$

$$= 0.0049 \, \mathrm{m} \, (4.9 \, \mathrm{mm}).$$

Exercises

1 Viscous flow takes place between two stationary, plane parallel plates, distance t apart, and the width in a direction transverse to the direction of flow is large compared with t, so that side effects may be neglected. Show that the pressure drop per unit length in the direction of flow is given by

$$\frac{p}{L} = \frac{12 \mu u_\mathrm{m}}{t^2}$$

where μ_m = the mean velocity of flow,
and μ = the coefficient of absolute viscosity.

A piston of diameter D and length L moves concentrically in an oil dashpot with velocity u_P. If F is the load on the piston, and the small radial clearance is t, show that an approximate expression for the velocity of the piston (of assumed negligible mass) is:

$$u_\mathrm{P} = \frac{4t^3 F}{3\pi D^3 L\mu}.$$

2 A long piston of diameter 75 mm, with its axis vertical, is partly engaged in a cylinder with a radial clearance of 0.075 mm, the remaining space in the cylinder being filled with oil. Under an axial load of 400 N, its penetration is increased from 50 mm to 100 mm in 30 seconds. Derive an approximate expression for the

velocity of the dashpot piston and use it to obtain the time taken to increase the piston penetration of the cylinder. Hence, find the viscosity of the oil in N s/m^2.

Assume $-(dp/dx) = \dfrac{12\mu u_\mathrm{m}}{t^2}$ where u_m = mean velocity of flow.

(*Answer:* 0.001 36 N s/m^2)

3 The ram of a compression machine having a diameter of 305 mm is operated by oil having an absolute viscosity of 0.15 N s/m^2. The oil is pumped through a pipe of diameter 6.35 mm and of length 4.57 m to the ram cylinder by means of a hand pump having a plunger diameter of 25.4 mm. If the ram is exerting a force of 100 kN when its speed is 0.11 $\dfrac{\mathrm{mm}}{\mathrm{s}}$, deduce the force at the plunger.

(*Answer:* 763.7 N)

4 Oil of specific gravity 0.9 is pumped through a horizontal pipe of diameter 200 mm and length 300 m at the rate of 100 m^3/h. It takes 9 kW of power to drive the pump which has an efficiency of 68 per cent. Determine the velocity at a radius of 50 mm from the pipe axis and hence the dynamic pressure which would be indicated by a small Pitot tube placed in that position. Assume streamline flow for which the velocity distribution across the pipe is given by

$$u = \frac{\Delta p}{\Delta L}\,\frac{1}{4\mu}\,(r^2 - y^2),$$

where u is the velocity at a radius y, r is the radius of the pipe, μ is the coefficient of absolute viscosity and Δp is the pressure drop along the pipe for a length ΔL.

(*Answer:* 1.33 m/s, 794 N/m^2)

10 The Steady, One-dimensional Flow of an Ideal Compressible Fluid

10.1 Introduction

In this chapter we study the flow of a fluid having a variable density in which temperature becomes a significant quantity and the subjects of fluid mechanics and thermodynamics become inseparable.

Since we will assume that the flow is always steady we will need to use the steady-flow energy equation (refer to Chapter 3 for the relationship between Bernoulli's equation and the full energy equation).

Thus for unit mass flow rate:

$$h_1 + \frac{u_1^2}{2} + {}_1q_2 = h_2 + \frac{u_2^2}{2} + {}_1w_2. \qquad \ldots (1)$$

Note that the potential energy term gz is omitted since it is negligible as the following example illustrates.

Consider air at 1 bar, 20 °C, changing to 5 bar, 200 °C.
$h_2 - h_1 = c_p (T_2 \quad T_1)$ for a perfect gas

$$= 1.005 \frac{kJ}{kg\ K} (200 - 20)\ K = 180.9 \frac{kJ}{kg} \quad \text{(using tables by Rogers and Mayhew)}.$$

Now for $g(z_2 - z_1)$ to be of significance in the energy equation, $z_2 - z_1$ must be comparable with $\dfrac{h_2 - h_1}{g}$,

i.e. $z_2 - z_1$ must be comparable with $\dfrac{180.9 \frac{kJ}{kg}}{9.81 \frac{m}{s^2}} \left[\dfrac{kg\ m}{N\ s^2} \right]$,

i.e. $z_2 - z_1$ must be 18 440 m,
which is beyond the scope of general engineering practice.

The fluids in this chapter will either be vapours or assumed to be perfect gases and the properties of both classes of fluid have been fully covered in Chapter 4 of

Work Out Thermodynamics in this series and are not repeated here. Reference must also be made to Rogers and Mayhew.*

10.2 Stagnation Properties

In dealing with the steady-flow energy equation we have to decide if kinetic energy is appreciable or not (i.e. $u^2/2$ compared with h, q and w).

Supposing that it is (as is often the case particularly in nozzles and turbo-machinery), we can conveniently recast the energy equation as follows:

$$h_{01} + {_1}q_2 = h_{02} + {_1}w_2, \qquad \qquad \ldots (2)$$

where

$$h_0 = h + \frac{u^2}{2} \quad \text{by definition.} \qquad \ldots (3)$$

h_0 is called the stagnation enthalpy and is that enthalpy achieved when the fluid is brought adiabatically to rest ($u = 0$). Note that the deceleration does not have to be reversible but must not result in a heat or work transfer to or from the fluid.

Example

A rocket engine in which the thrust = 450 kN and the jet velocity = 3000 m/s.

$$\frac{u^2}{2} = \frac{3000}{2} \frac{m^2}{s^2} \left[\frac{N\,s^2}{kg\,m} \right] \left[\frac{kJ}{kN\,m} \right] = 4500 \frac{kJ}{kg}.$$

For such an engine a typical exhaust temperature would be 1200 K and c_p = 6.4 kJ/kg K for the products at this elevated temperature. h above datum = $c_p T$ = 6.4 kg/kg K × 1200 K = 7680 kJ/kg.

Clearly $u^2/2$ and h are comparable. (Sometimes $u^2/2 > h$.)

The deceleration process may be imagined to occur in a diffuser in which the flow is adiabatic as shown in the diagram. The enthalpy at diffuser exit where $u = 0$ is the stagnation value appropriate to that point in the actual stream.

We are often involved with gas dynamics since in many engineering problems the vapours are highly superheated at low pressure and behave like gases.

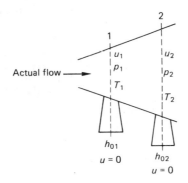

Restriction   If the temperature range involved in the problem is small then we may be permitted to use a constant mean value for the isobaric specific heat capacity c_p. Thus we can recast as follows:

$$c_p T_0 = c_p T + \frac{u^2}{2} \qquad \qquad \ldots (4)$$

*See footnote, page 95.

or
$$T_0 = T + \frac{u^2}{2c_p}.$$
...(5)

T_0 is called the stagnation temperature and is the sum of T the static temperature and $u^2/2c_p$ the dynamic temperature.

The words 'static' and 'stagnation' cause confusion in the student's mind since they both suggest zero kinetic energy but actually only T_0 relates to this state. T is the temperature measured in a flowing stream of fluid by an instrument moving with the stream, or without appreciable fluid deceleration due to the introduction of a temperature-measuring instrument.

Note that $u^2/2c_p$ can be written as θ_u (the temperature equivalent of u).

Pressure   The relationship between T and T_0 is an energy relationship and viscous friction in the fluid does not change the resultant value of T_0 when the fluid decelerates to rest (in the absence of heat and work transfer).

However, friction does affect momentum and in order to realise the full value of stagnation pressure p_0 from the static value p we have to place a further restriction of reversibility on the flow in the imaginary diffuser. That is, for the full p_0 to be achieved the flow in the diffuser must be both adiabatic and reversible, i.e. isentropic (s = constant).

Then
$$\frac{p_0}{p} = \left(\frac{T_0}{T}\right)^{\gamma/(\gamma-1)} \qquad \text{for a gas,}$$

or p_0 relates to p with s = constant (using tables appropriately for a vapour).

Summarising, we see that in a duct with adiabatic flow and no work, stagnation temperature will be constant whether the flow is reversible or not, but p_0 will decrease in irreversible flow and only remains constant in reversible flow.

Acoustic Velocity and Mach Number

Propagation Velocity of a Plane Pressure Pulse of Small Amplitude

Diagrams (a) and (b) depict diagrammatically the flow of a small-amplitude pressure wave in a motionless fluid in a duct.

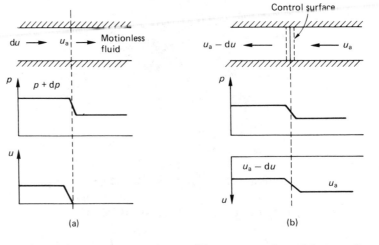

(a)

Observer at rest.

(b)

Observer moving with wave front (equivalent to superposition of a leftward velocity u_a on the flow shown in (a)).

Momentum: Shear forces on the control surface are very small compared with pressure forces.

$$A \left[p - (p + dp) \right] = \dot{m} \left[(u_a - du) - u_a \right].$$

Mass:
$$\dot{m} = \rho u_a A.$$

Thus
$$dp = \rho u_a \, du. \qquad \ldots \text{(A)}$$

(Euler's equation; du is decrease in direction of motion.)

For an unchanged area on both sides of the wavefront,

$$\rho u_a = (\rho + d\rho)(u_a - du)$$

or
$$\frac{d\rho}{\rho} = \frac{du}{u_a}, \qquad \ldots \text{(B)}$$

and from (A) and (B),
$$u_a^2 = \left(\frac{dp}{d\rho} \right) = \left(\frac{\partial p}{\partial \rho} \right)_s. \qquad \ldots \text{(C)}$$

The justification for this step is that acoustic waves have small amplitude and minimal effect on the flow, and thus the latter may be assumed to be reversible. Since the flow is at high speed it is also assumed to be adiabatic and thus $s = \text{con-stant}$.

Perfect gas:
$$\frac{p}{\rho^\gamma} = \text{constant k}.$$

Thus
$$\ln p = \ln k + \ln \rho^\gamma$$

or
$$\ln p - \gamma \ln \rho = \text{constant}$$

and
$$\frac{dp}{p} = \frac{\gamma \, d\rho}{\rho}.$$

Thus
$$\left(\frac{\partial p}{\partial \rho} \right)_s = \frac{\gamma p}{\rho} = \gamma R T \qquad (p = \rho R T \text{ for a gas})$$

i.e.
$$u_a = \sqrt{\frac{\gamma p}{\rho}} = \sqrt{\gamma R T} = \sqrt{\gamma \, \frac{R_0}{m_v} T}, \qquad \ldots \text{(D)}$$

where m_v = molecular mass and R_0 = universal gas constant.

Mach number (Ma): The flow pattern depends on the magnitude of u and u_a where u is the fluid particle velocity.

For $(Ma) < 1$: subsonic flow,
$(Ma) = 1$: sonic flow,
$(Ma) > 1$: supersonic flow.

Flow relationships for a gas: For all adiabatic flow (reversible or irreversible),

$$T_0 = T + \frac{u^2}{2 c_p} = T + \frac{u^2}{2 \left(\frac{\gamma}{\gamma - 1} \right) R} = T + \frac{(Ma)^2 \left(\frac{\gamma p}{\rho} \right)}{2 \left(\frac{\gamma}{\gamma - 1} \right) R} \qquad \text{from (D)}$$

$$= T \left[1 + \left(\frac{\gamma - 1}{2} \right) (Ma)^2 \right]. \qquad \ldots \text{(E)}$$

143

For reversible adiabatic, i.e. isentropic flow only:

$$\frac{p_0}{p} = \left(\frac{T_0}{T}\right)^{\gamma/(\gamma-1)} = \left[1 + \frac{\gamma-1}{2}\,(Ma)^2\right]^{\gamma/(\gamma-1)}, \qquad \ldots \text{(F)}$$

$$\frac{\rho_0}{\rho} = \frac{p_0\,RT}{p\,RT_0} = \left[1 + \left(\frac{\gamma-1}{2}\right)(Ma)^2\right]^{\gamma/(\gamma-1)}\left[1 + \left(\frac{\gamma-1}{2}\right)(Ma)^2\right]^{-1}$$

$$= \left[1 + \left(\frac{\gamma-1}{2}\right)(Ma)^2\right]^{1/(\gamma-1)} \qquad \ldots \text{(G)}$$

At critical (or sonic) conditions, $(Ma) = 1$ $\qquad (u = u^*)$

and $\dfrac{T^*}{T_0} = \dfrac{2}{\gamma+1}$; $\qquad \dfrac{p^*}{p_0} = \left(\dfrac{2}{\gamma+1}\right)^{\gamma/(\gamma-1)}$; $\qquad \dfrac{\rho^*}{\rho_0} = \left(\dfrac{2}{\gamma+1}\right)^{1/(\gamma-1)}$ $\quad \ldots \text{(H)}$

$$\dot{m} = \rho u A,$$

$$\frac{\dot{m}}{A} = \rho u = \frac{p}{RT}\,u = \frac{pu}{\sqrt{\gamma R T}}\sqrt{\frac{\gamma}{R}}\sqrt{\frac{T_0}{T}}\frac{1}{\sqrt{T_0}}$$

$$= \sqrt{\frac{\gamma}{R}}\frac{p}{\sqrt{T_0}}\,(Ma)\sqrt{1 + \left(\frac{\gamma-1}{2}\right)(Ma)^2}$$

and from previous relations for $\dfrac{p}{p_0}$, (Ma)

$$\frac{\dot{m}}{A} = \sqrt{\frac{\gamma}{R}}\frac{p_0}{1 + \left(\frac{\gamma-1}{2}\right)(Ma)^2}\frac{1}{\sqrt{T_0}}\,(Ma)\sqrt{1 + \left(\frac{\gamma-1}{2}\right)(Ma)^2}$$

$$= \sqrt{\frac{\gamma}{R}}\frac{p_0}{\sqrt{T_0}}\frac{(Ma)}{\left[1 + \left(\frac{\gamma-1}{2}\right)(Ma)^{(\gamma+1)/2(\gamma-1)}\right]} \qquad \ldots \text{(I)}$$

for $(Ma) = 1$ $\quad\left(\dfrac{\dot{m}}{A}\right)_{max} = \sqrt{\dfrac{\gamma}{R}\left(\dfrac{2}{\gamma+1}\right)^{(\gamma+1)/(\gamma-1)}}\dfrac{p_0}{\sqrt{T_0}}$. $\qquad \ldots \text{(J)}$

Defining specific impulse $I = \dfrac{\text{net thrust}}{\text{mass flow rate}} = \dfrac{F_N}{\dot{m}}$, it can be shown that

$$\frac{I}{I^*} = \frac{1 + \gamma\,(Ma)^2}{(Ma)\sqrt{2\,(\gamma+1)\left[1 + \left(\frac{\gamma-1}{2}\right)(Ma)^2\right]}} . \qquad \ldots \text{(K)}$$

Values of all the quantities listed in equations (A) to (K) can be found in compressible flow tables for air ($\gamma = 1.4$) in the range $0 < (Ma) < 10$.

Conditions Inside a Nozzle or Diffuser

Assuming isentropic flow from state 1 to state 2,

$$h = e + pv \qquad \text{(definition)}$$

$$dh = de + p\,dv + v\,dp.$$

First law of thermodynamics:

$$dq_R = de + p\,dv \qquad (R \equiv \text{reversible})$$

and for isentropic flow
$$dq_R = 0.$$

Thus
$$dh = v\,dp$$

and, integrating,
$$h_2 - h_1 = \int_1^2 v\,dp$$

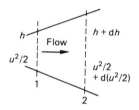

Steady-flow energy equation:

$$h + \frac{u^2}{2} = h + dh + \frac{u^2}{2} + d\left(\frac{u^2}{2}\right),$$

$$d\left(\frac{u^2}{2}\right) = -dh = -v\,dp,$$

$$u\,du = -v\,dp,$$

$$\frac{du}{u} = \frac{v\,dp}{u^2}. \qquad \ldots (L)$$

Mass continuity: $\quad \dot{m} = \dfrac{Au}{v} = $ a constant, and, taking logs,

$$\log A + \log u - \log v = \log \text{constant} = \text{constant}$$

Differentiating, $\qquad \dfrac{dA}{A} + \dfrac{du}{u} - \dfrac{dv}{v} = 0. \qquad \ldots (M)$

Bulk modulus of fluid $K = \dfrac{-dp}{dv/v} \qquad$ (see Chapter 1).

Thus $\qquad \dfrac{-dv}{v} = \dfrac{dp}{K}, \qquad$ and, substituting in M,

$$\frac{dA}{A} = \left[\frac{v}{u^2} - \frac{1}{K}\right] dp. \qquad \ldots (N)$$

Criteria for Accelerating and Decelerating Flow

Case 1

Negative dp: $\qquad$ dA is positive when $u^2 > vK$,

i.e. $\qquad$ dA is positive when $u^2 > \dfrac{K}{\rho}$.

Now $$pv^\gamma = \text{constant } C,$$

and $$K = -v\,\frac{\mathrm{d}p}{\mathrm{d}v} = -C(-\gamma)v^{-\gamma} = C\gamma v^{-\gamma} = \gamma p.$$

Thus $$\frac{K}{\rho} = \frac{\gamma p}{\rho} \qquad \text{and} \qquad \sqrt{\frac{K}{\rho}} = \sqrt{\frac{\gamma p}{\rho}} = u_a \text{ (acoustic velocity)}.$$

Thus $\mathrm{d}A$ is positive when $u > u_a$ (i.e. $(Ma) > 1$).

Thus duct diverges.

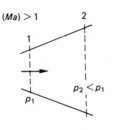

Supersonic flow: duct diverges for p decreasing.

Case 2

Conversely, for negative $\mathrm{d}p$ and $u^2 < vk$, $\mathrm{d}A$ negative,

$$u < u_a \qquad \text{(subsonic flow)} \qquad ((Ma) < 1).$$

Thus duct converges.

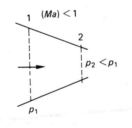

Subsonic flow: duct converges for p decreasing.

Case 3

Positive $\mathrm{d}p$, $\mathrm{d}A$ is +ve when $u < u_a$ $\qquad ((Ma) < 1)$.
Subsonic flow, duct diverges for pressure increasing.

Case 4

Positive $\mathrm{d}p$, $\mathrm{d}A$ is −ve when $u > u_a$ $\qquad ((Ma) > 1)$.
Supersonic flow, duct converges for pressure increasing.
Thus for continuous expansion of a fluid inside a nozzle the duct walls must converge till the fluid velocity reaches the sonic threshold $((Ma) = 1)$ and then diverge for further expansion.

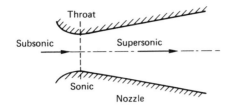

p decreases continuously left to right.

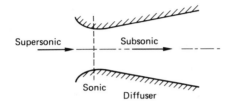

p increases continuously left to right.

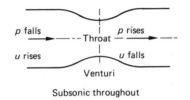

10.4 Critical Pressure Ratio in a Nozzle

$$\left(\frac{p_t}{p_1}\right)^* = \left(\frac{2}{\gamma + 1}\right)^{n/(n-1)}$$

A standard proof (omitted here).

n	1.135	1.3	1.4	1.667
$\left(\frac{p_t}{p}\right)^*$	0.577	0.546	0.528	0.487

$n = 1.135$ for wet steam.
$n = 1.3$ for superheated steam.
$n = \gamma = 1.4$ for air.

10.5 Nozzle Efficiency (η_N)

Assume that u_1 is significant.
In a nozzle there is no heat or work transfer.
Steady-flow energy equation:

$$h_1 + \frac{u_1^2}{2} = h_2 + \frac{u_2^2}{2} \qquad \text{(irreversible case)}$$

$$= h_{2s} + \frac{u_{2s}^2}{2} \qquad \text{(reversible, i.e. isentropic, case).}$$

147

Thus
$$h_{01} = h_2 + \frac{u_2^2}{2} = h_{2s} + \frac{u_{2s}^2}{2}$$

and
$$\eta_N = \frac{u_2^2}{u_{2s}^2} \qquad \text{(definition)}.$$

Note that this results in $\quad \eta_N = \dfrac{h_{01} - h_2}{h_{02} - h_{2s}}$

and if u_1 is small compared with h_1, $\qquad\qquad\qquad\qquad$ ... (O)

$$\eta_N = \frac{h_1 - h_2}{h_1 - h_{2s}}.$$

Nozzle efficiency calculation

(a) Vapour: equate $s_1 = s_2$, find h_{2s} and hence η_N.

(b) Gas: use $T_{2s} = T_1 \left(\dfrac{p_2}{p_1}\right)^{(\gamma - 1/\gamma)}$ and then

$$h_{2s} - h_1 = c_p \, (T_{2s} - T_1) \text{ and hence } \eta_N.$$

10.6 Worked Examples

Note: In the preceding chapters of this book the pressure of the fluid has been assumed to be a gauge pressure. In this chapter it should be borne in mind that all pressures are absolute in value.

Example 10.1

Steam flows through a duct of diameter 15 cm. The duct is fed from a large insulated rigid vessel where the state of the steam is maintained at 7 bar and 200 °C.

At a certain cross-section in the duct, the state of the steam is 4 bar and 0.95 dry. The stagnation pressure at this section is 5 bar.

Determine the steam velocity at this section of the duct and the rate of heat transfer through the duct walls.

Answer

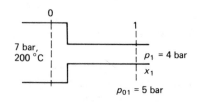

Properties of steam are taken from tables by Rogers and Mayhew.

Page 7 (at 7 bar, 200 °C): $\quad h_{00} = 2846 \, \dfrac{\text{kJ}}{\text{kg}}$.

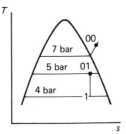

Page 4 (at 4 bar, 0.95 dry): $h_1 = h_{f,1} + x_1 h_{fg,1}$

$$= 605 + 0.95 (2134) = 2632.3 \frac{kJ}{kg}.$$

$$s_1 = s_{f,1} + x s_{fg,1} = 1.776 + 0.95 (5.121)$$

$$= 6.641 \frac{kJ}{K\,kg} = s_{01} \qquad (u_1 = 0).$$

$$x_{01} = \frac{s_{01} - s_{f,1}}{s_{fg,1}} = \frac{6.641 - 1.96}{4.962} = 0.964.$$

$$h_{01} = h_{f,01} + x_{01} h_{fg,01} = 640 + 0.964 (2109)$$

$$= 2672.1 \frac{kJ}{kg}.$$

Steady-flow energy equation $(0 \to 1)$:

$$\dot{m} (h_{00} - h_{01}) = {}_0\dot{Q}_1 \qquad \text{where } {}_0\dot{Q}_1 \text{ is heat rejected.}$$

Now $u_1 = \sqrt{2 (h_{01} - h_1)} = \sqrt{2 (2672.1 - 2632.3) \frac{kJ}{kg} \left[\frac{kg\,m}{N\,s^2} \right] \left[\frac{10^3\,N\,m}{kJ} \right]}$

$$= 282 \frac{m}{s}.$$

Mass continuity: $\dot{m} = \dfrac{A_1 u_1}{v_1} = \dfrac{A_1 u_1}{x_1 v_{g,1}} = \dfrac{\pi \times 0.15^2\ m^2 \times 282\ \frac{m}{s}\ kg}{4 \times 0.95 \times 0.4623\ m^3}$

$$= 11.35 \frac{kg}{s} \qquad \text{since } v_{g,1} = 0.4623 \frac{m^3}{kg} \quad \text{(page 4).}$$

$${}_0\dot{Q}_1 = 11.35 \frac{kg}{s} (2846 - 2672.1) \frac{kJ}{kg} = 1973.8 \text{ kW.}$$

Example 10.2

The following data refer to the test on an axial-flow compressor. Atmospheric temperature and pressure at inlet are 15.5 °C and 10^5 N/m², respectively. Stagnation temperature and pressure in the delivery pipe are 160 °C and 3.4×10^5 N/m² respectively.

The static pressure in the delivery pipe is 2.9×10^5 N/m².

Calculate (a) the stagnation isentropic efficiency of the machine,

(b) the air velocity in the delivery pipe.

Answer

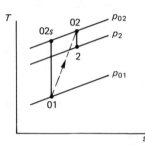

Air is assumed to be a perfect gas and the temperature–entropy diagram is given above.

$$T_{02s} = T_{01}\left(\frac{p_{02}}{p_{01}}\right)^{(\gamma-1)/\gamma} = (273 + 15.5)\ \text{K}\ (3.4)^{0.286} = 409.4\ \text{K}$$

since γ for air = 1.4 (page 24 of tables) and $\dfrac{\gamma - 1}{\gamma} = \dfrac{0.4}{1.4} = 0.286$.

The stagnation isentropic efficiency of the machine is defined as

$$\eta_0 = \frac{w_{\text{ideal}}}{w_{\text{actual}}} = \frac{h_{02s} - h_{01}}{h_{02} - h_{01}} \qquad \text{(in the absence of heat transfer).}$$

and for a perfect gas for which c_p is constant,

$$\eta_0 = \frac{T_{02s} - T_{01}}{T_{02} - T_{01}} = \frac{409.4 - 288.5}{160 - 15.5} = 0.837.$$

(Note that when subtracting two temperatures it is not necessary to add 273 to each to make them absolute since these cancel out.)

$$T_2 = T_{02}\left(\frac{p_2}{p_{02}}\right)^{(\gamma-1)/\gamma} = (273 + 160)\ \text{K}\left(\frac{2.9}{3.4}\right)^{0.286} = 413.7\ \text{K}$$

$$u_2 = \sqrt{2c_p(T_{02} - T_2)} = \sqrt{2 \times 1.005\ \frac{\text{kJ}}{\text{kg K}}(433 - 413.7)\ \text{K}\left[\frac{\text{kg m}}{\text{N s}^2}\right]\left[\frac{10^3\ \text{N m}}{\text{kJ}}\right]}$$

$$= 197\ \frac{\text{m}}{\text{s}} \qquad \left(\text{where } c_p = 1.005\ \frac{\text{kJ}}{\text{kg K}} \text{ for air; page 24}\right).$$

Example 10.3

A venturimeter used to measure the flow of air has a throat diameter of 70 mm and an isentropic efficiency of 0.97. Determine the mass flow rate in kg/s when the inlet pressure and temperature are 1.25 bar and 15 °C respectively and the inlet velocity is small. The throat pressure is 1.1 bar and the properties of air may be taken from tables page 24.

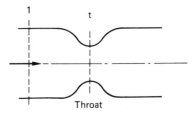

Throat

$$T_{ts} = T_1\left(\frac{p_t}{p_1}\right)^{(\gamma-1)/\gamma} = (15 + 273)\,\text{K}\left(\frac{1.1}{1.25}\right)^{0.286} = 277.7\,\text{K},$$

(γ, c_p and R for air are on page 24 of tables)

$$T_1 - T_t = \eta\,(T_1 - T_{ts}) \qquad (c_p\text{s cancel}).$$

Thus $T_1 = T_t = 0.97\,(288 - 277.7) = 10\,\text{K}$

$$= 288 - 10 = 278\,\text{K}.$$

$$\rho_t = \frac{p_t}{RT_t} = \frac{110\,\dfrac{\text{kN}}{\text{m}^2}\left[\dfrac{\text{J}}{\text{N m}}\right]}{0.287\,\dfrac{\text{kJ}}{\text{kg K}} \times 278\,\text{K}} = 1.379\,\frac{\text{kg}}{\text{m}^3}$$

$$\frac{u_t^2}{2} = c_p\,(T_1 - T_t) = \left(\frac{\gamma}{\gamma-1}\right)R\,(T_1 - T_t)$$

$$u_t = \sqrt{2 \times 3.5 \times 0.287\,\frac{\text{kJ}}{\text{kg K}}\,(10\,\text{K})\cdot\left[\frac{\text{kg m}}{\text{N s}^2}\right]\left[\frac{10^3\,\text{N m}}{\text{kJ}}\right]}$$

$$= 141.7\,\frac{\text{m}}{\text{s}}.$$

Mass flow rate $= \dot{m} = \rho_t A_t u_t$

$$= 1.379\,\frac{\text{kg}}{\text{m}^3} \times \frac{\pi}{4} \times 0.07^2\,\text{m}^2 \times 141.7\,\frac{\text{m}}{\text{s}}$$

$$= 0.752\,\frac{\text{kg}}{\text{s}}.$$

Example 10.4

A pitot tube in an air duct indicates a stagnation pressure of 700 mmHg when the static pressure in the duct is 500 mmHg. A temperature probe shows a stagnation temperature of 65 °C. What is the air velocity and Mach number upstream of the pitot tube? Use air properties given on page 24 of tables.

Answer

$$p_{01} = 0.7\,\text{mHg}\left[13\,600\,\frac{\text{kg}}{\text{m}^3}\right]9.81\,\frac{\text{m}}{\text{s}^2}\left[\frac{\text{N s}^2}{\text{kg m}}\right] \qquad (\text{i.e. } p = \rho g H)$$

151

$$= 93.39 \ \frac{\text{kN}}{\text{m}^2}.$$

$$p_1 = \frac{500}{700} \times 93.39 = 66.7 \ \frac{\text{kN}}{\text{m}^2}.$$

$$T_{01} = 65 + 273 = 338 \text{ K}.$$

$$T_1 = T_{01}\left(\frac{p_1}{p_{01}}\right)^{(\gamma-1)/\gamma} = 338 \text{ K} \left(\frac{66.7}{93.39}\right)^{0.286} = 307 \text{ K}.$$

$$u_1 = \sqrt{2c_p \, (T_{01} - T_1)} = \sqrt{2\left(\frac{\gamma}{\gamma - 1}\right) R \, (T_{01} - T_1)}$$

$$= \sqrt{2 \times 3.5 \times 0.287 \ \frac{\text{kJ}}{\text{kg K}} (338 - 307) \text{ K} \left[\frac{\text{kg m}}{\text{N s}^2}\right] \left[\frac{10^3 \text{ N m}}{\text{kJ}}\right]}$$

$$= 249.6 \ \frac{\text{m}}{\text{s}}.$$

$$(Ma) = \frac{u}{u_a} = \frac{u}{\sqrt{\gamma R T}} = \frac{249.6 \ \dfrac{\text{m}}{\text{s}}}{\sqrt{1.4 \times 0.287 \ \dfrac{\text{kJ}}{\text{kg K}} \times 307 \text{ K} \left[\dfrac{\text{kg m}}{\text{N s}}\right] \left[\dfrac{10^3 \text{ N m}}{\text{kJ}}\right]}}$$

$$= \frac{249.6}{351.2} = 0.71.$$

Example 10.5

A large air-vessel is maintained at a pressure of 630 kN/m² and a temperature of 15 °C. It is connected by a convergent nozzle with a throat area of 75×10^{-6} m² to another vessel of capacity 0.75 m³ containing air initially at 105 kN/m² and 22 °C. The nozzle is opened until conditions in the second vessel become 315 kN/m² and 18 °C. Determine the period of time the nozzle is open. The properties of air may be taken from page 24 of tables.

Answer

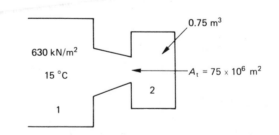

Assume isentropic flow in the nozzle.

$$\text{Critical pressure ratio} = \left(\frac{2}{\gamma + 1}\right)^{\gamma/(\gamma-1)} = \left(\frac{2}{2.4}\right)^{3.5} = 0.528.$$

Thus there is always sonic flow at the throat since

$$p_2 < p_t^* \qquad \left(p_t^* = 0.528 \times 630 = 332.8 \ \frac{\text{kN}}{\text{m}^2}\right).$$

$$T_t = T_{01}\left(\frac{p_t}{p_{01}}\right)^{(\gamma-1)/\gamma} = T_{01}\left(\frac{2}{\gamma+1}\right) = \frac{288\ K \times 2}{2.4} = 240\ K.$$

$$u_t = \sqrt{2c_p(T_0 - T_t)} = \sqrt{2\left(\frac{\gamma}{\gamma-1}\right)R(T_0 - T_t)}$$

$$= \sqrt{2 \times 3.5 \times 0.287\ \frac{kJ}{kg\ K}\ (288 - 240)\ K\ \left[\frac{kg\ m}{N\ s^2}\right]\left[\frac{10^3\ N\ m}{kJ}\right]} = 310.5\ \frac{m}{s}.$$

$$\dot{m} = \rho_t A_t u_t = \frac{p_t}{RT_t}\ u_t A_t = \frac{\frac{75}{10^6}\ m^2 \times 310.5\ \frac{m}{s} \times 332.8\ \frac{kN}{m^2}}{0.287\ \frac{kJ}{kg\ K} \times 240\ K}$$

$$= 0.1125\ \frac{kg}{s}.$$

$$m_1\ \text{initially in second vessel} = \frac{p_1 V}{RT_1} = \frac{105\ \frac{kN}{m^2} \times 0.75\ m^3}{0.287\ \frac{kJ}{kg\ K} \times 295\ K} = 0.93\ kg;$$

$$m_2\ \text{finally in second vessel} = \frac{p_2 V}{RT_2} = \frac{315 \times 0.75}{0.287 \times 291} = 2.829\ kg.$$

$$\text{Time } t = \frac{m_2 - m_1}{\dot{m}} = \frac{(2.829 - 0.93)\ kg}{0.1125\ \frac{kg}{s}} = 16.9\ s.$$

Example 10.6

Gas approaches a convergent–divergent nozzle at a pressure of 5 bar and a temperature of 500 °C but with negligible velocity.

The area of the throat is 100 m².

Flow between inlet and throat may be assumed isentropic but the overall nozzle isentropic efficiency is 0.95. The nozzle discharges at a pressure of 1 bar.

Calculate:

(a) the gas velocity at the throat,
(b) the gas velocity at exit,
(c) the exit area of the nozzle,
(d) the mass flow rate.

Take $\gamma = 1.4$ and $c_p = 1.04$ kJ/kg K.

Answer

$$p_t = \left(\frac{2}{\gamma+1}\right)^{\gamma/(\gamma-1)} p_1 = \left(\frac{2}{2.4}\right)^{3.5} \times 5\ bar = 2.64\ bar,$$

$$T_t = T_1\left(\frac{2}{\gamma+1}\right) = \frac{(500 + 273)\ 2}{2.4} = 644.2\ K,$$

$$R = \frac{\gamma-1}{\gamma}\ c_p = 0.286 \times 1.04\ \frac{kJ}{kg\ K} = 0.297\ \frac{kJ}{kg\ K},$$

$$u_t = \sqrt{\gamma RT} = \sqrt{1.4 \times 0.297\ \frac{kJ}{kg\ K} \times 644.2\ K\ \left[\frac{kg\ m}{N\ s^2}\right]\left[\frac{10^3\ N\ m}{kJ}\right]}$$

$$= 518.3\ \frac{m}{s}.\quad (a)$$

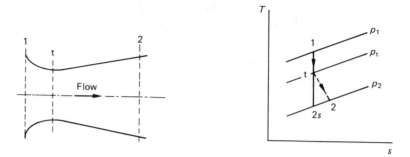

$$T_{2s} = T_1\left(\frac{p_2}{p_1}\right)^{(\gamma-1)/\gamma} = 773 \text{ K} \left(\frac{1}{5}\right)^{0.286} = 487.8 \text{ K},$$

$$T_1 - T_{2s} = 773 - 487.8 = 285.2 \text{ K},$$

$$T_1 - T_2 = \eta\,(T_1 - T_{2s}) = 0.95 \times 285.2 \text{ K} = 270.9 \text{ K},$$

$$T_2 = 502.1 \text{ K}.$$

$$u_2 = \sqrt{2c_p\,(T_1 - T_2)} = \sqrt{1.4 \times 0.298 \times 644 \times 1000} = 750.6 \ \frac{\text{m}}{\text{s}}. \quad \text{(b)}$$

$$v_c = \frac{RT_c}{p_t} = \frac{0.297 \ \frac{\text{kJ}}{\text{kg K}} \times 644.2 \text{ K}}{264 \ \frac{\text{kN}}{\text{m}^2}} = 0.725 \ \frac{\text{m}^3}{\text{kg}},$$

$$\dot{m} = \frac{u_t A_t}{v_t} = \frac{518.3 \ \frac{\text{m}}{\text{s}} \times 0.0001 \text{ m}^2}{0.725 \ \frac{\text{m}^3}{\text{kg}}} = 0.0715 \ \frac{\text{kg}}{\text{s}}. \quad \text{(d)}$$

$$A_2 = \frac{\dot{m}v_2}{u_2} = \frac{\dot{m}RT_2}{u_2 p_2} = \frac{0.0715 \ \frac{\text{kg}}{\text{s}} \times 0.298 \ \frac{\text{kJ}}{\text{kg K}} \times 502.1 \text{ K}}{750.6 \ \frac{\text{m}}{\text{s}} \times 100 \ \frac{\text{kN}}{\text{m}^2}} \left[\frac{10^6 \text{ mm}^2}{\text{m}^2}\right]$$

$$= 143 \text{ mm}^2. \quad \text{(c)}$$

Example 10.7

A convergent nozzle is supplied with steam at static conditions of 1.8 bar and 150 °C and a velocity of 205 m/s.

Calculate the exit pressure and mass flow rate per unit exit area for maximum discharge conditions assuming isentropic flow.

Take the critical pressure ratio as 0.55.

Answer

$$h_{01} = h_1 + \tfrac{1}{2}u_1^2 = \left[2773 + \frac{0.3}{0.5}\,(2770 - 2773)\right]\frac{\text{kJ}}{\text{kg}} + \frac{205^2}{2}\frac{\text{m}^2}{\text{s}^2}\left[\frac{\text{N s}^2}{\text{kg m}}\right]\left[\frac{\text{kJ}}{10^3 \text{ N m}}\right]$$

(interpolating for h_1 at 1.8 bar on page 6 of tables)

$$= 2795.8 \ \frac{\text{kJ}}{\text{kg}}.$$

154

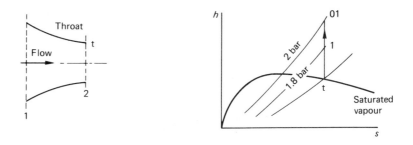

The easiest way to find p_{01} is to use the enthalpy–entropy chart. Moving vertically upwards from 1.8 bar, 150°C to $h_{01} = 2795.8$ (kJ/kg) gives $p_{01} = 2$ bar (approximately).

Then $\quad p_{\text{exit}} = p_t = \left(\dfrac{2}{n+1}\right)^{n/(n-1)} p_{01} = 0.546 \times 2 = 1.092$ bar.

Thus coming vertically down to 1.092 bar from 1.8 bar, 150°C gives saturated steam at throat (or near enough).

Then $h_2 = h_t = 2675 + \dfrac{92}{100}\,(2680 - 2675) = 2679.6\ \dfrac{\text{kJ}}{\text{kg}}$ $\qquad$ (page 4 of tables),

$v_2 = v_t = 1.694 + \dfrac{92}{100}\,(1.594 - 1.694) = 1.5523\ \dfrac{\text{m}^3}{\text{kg}}$ $\qquad$ (page 4 of tables),

$u_2 = \sqrt{2\,(h_{01} - h_2)} = \sqrt{2\,(2795.8 - 2679.6)}\ \dfrac{\text{kJ}}{\text{kg}}\left[\dfrac{\text{kg m}}{\text{N s}^2}\right]\left[\dfrac{10^3\ \text{N m}}{\text{kJ}}\right]$

$\qquad = 482.1\ \dfrac{\text{m}}{\text{s}}\ ,$

and $\quad \dfrac{\dot{m}}{A_2} = \dfrac{u_2}{v_2} = \dfrac{482.1\ \dfrac{\text{m}}{\text{s}}}{1.5523\ \dfrac{\text{m}^3}{\text{kg}}} = 310.6\ \dfrac{\text{kg}}{\text{m}^2\,\text{s}}\ .$

Example 10.8

An air nozzle is fed by a pipe of internal diameter 50 mm in which the fluid pressure and temperature are maintained at 700 kN/m² and 200°C respectively. The nozzle back pressure is 1 bar.

Determine the throat and exit areas to give a mass discharge rate of 1 kg/s, assuming isentropic expansion to the throat and an overall isentropic efficiency of 0.85.

The critical pressure ratio for is 0.528.

Answer

Mass continuity: $u_1 = \dfrac{\dot{m}v_1}{A_1} = \dfrac{\dot{m}RT_1}{p_1 A_1} = \dfrac{1\ \dfrac{\text{kg}}{\text{s}} \times 0.287\ \dfrac{\text{kJ}}{\text{kg K}} \times 473\ \text{K}}{700\ \dfrac{\text{kN}}{\text{m}^2} \times \dfrac{\pi}{4} \times 0.05^2\ \text{m}^2}$

$\qquad\qquad = 98.8\ \dfrac{\text{m}}{\text{s}}\ .$

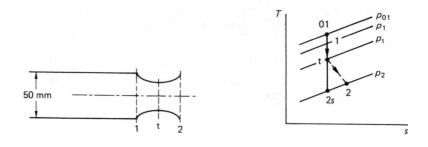

$$T_{01} = T_1 + \theta_{u,1} = 473 \text{ K} + \cfrac{98.9^2 \ \cfrac{\text{m}^2}{\text{s}^2}}{2 \times 1.005 \ \cfrac{\text{kJ}}{\text{kg K}} \left[\cfrac{\text{kg m}}{\text{N s}^2}\right] \left[\cfrac{10^3 \text{ N m}}{\text{kJ}}\right]}$$

$$= 473 \text{ K} + 4.9 \text{ K} = 477.9 \text{ K}.$$

$$p_{01} = p_1 \left(\frac{T_{01}}{T_1}\right)^{\gamma/(\gamma-1)} = 700 \ \frac{\text{kN}}{\text{m}^2} \left(\frac{477.9}{473}\right)^{3.5} = 725.5 \ \frac{\text{kN}}{\text{m}^2}.$$

$$p_t = p_{01} \left(\frac{2}{\gamma+1}\right)^{\gamma/(\gamma-1)} = 725.5 \ \frac{\text{kN}}{\text{m}^2} \times 0.528 = 383.3 \ \frac{\text{kN}}{\text{m}^2},$$

$$T_t = T_{01} \left(\frac{2}{\gamma+1}\right) = 477.9 \text{ K} \left(\frac{2}{2.4}\right) = 398.3 \text{ K},$$

$$u_t = u_{at} = \sqrt{\gamma R T_t} = \sqrt{1.4 \times 0.287 \ \frac{\text{kJ}}{\text{kg K}} \times 398.3 \text{ K} \left[\frac{\text{kg m}}{\text{N s}^2}\right] \left[\frac{10^3 \text{ N m}}{\text{kJ}}\right]} = 400 \ \frac{\text{m}}{\text{s}}.$$

$$v_t = \frac{R T_t}{p_t} = \cfrac{0.287 \ \cfrac{\text{kJ}}{\text{kg K}} \times 398.3 \text{ K}}{383.3 \ \cfrac{\text{kN}}{\text{m}^2}} = 0.298 \ \frac{\text{m}^3}{\text{kg}},$$

$$A_t = \frac{\dot{m} v_t}{u_t} = \cfrac{1 \ \cfrac{\text{kg}}{\text{s}} \times 0.298 \ \cfrac{\text{m}^3}{\text{kg}}}{400 \ \cfrac{\text{m}}{\text{s}}} \left[\frac{10^6 \text{ mm}^2}{\text{m}^2}\right] = 745.6 \text{ mm}^2.$$

$$T_{2s} = T_{01} \left(\frac{p_2}{p_{01}}\right)^{(\gamma-1)/\gamma} = 477.9 \text{ K} \left(\frac{100}{725.5}\right)^{0.286} = 271.1 \text{ K},$$

$$T_{01} - T_{2s} = 206.8 \text{ K},$$

$$T_{01} - T_2 = \eta (T_{01} - T_{2s}) = 0.85 \times 206.8 = 175.7 \text{ K},$$

$$T_2 = 302.2 \text{ K}.$$

$$v_2 = \frac{R T_2}{p_2} = \frac{0.287 \times 302.2}{100} = 0.867 \ \frac{\text{m}^3}{\text{kg}},$$

$$u_2 = \sqrt{2(h_{01} - h_2)} = \sqrt{2(c_p)(T_{01} - T_2)} = \sqrt{2 \times 1.005 \times 175.7 \times 1000}$$

$$= 594.3 \ \frac{\text{m}}{\text{s}},$$

$$A_2 = \frac{\dot{m} v_2}{u_2} = \frac{1 \times 0.867}{594.3} \times 10^6 = 1459 \text{ mm}^2 \quad \text{(as in previous part for dimensions)}.$$

Example 10.9

Gases leave the turbine of a jet propulsion unit at a pressure of 158.5 kN/m² and a temperature of 607.7 °C with a velocity of 400 m/s. The gases then enter the expansion nozzle where they are expanded adiabatically to 500 °C.

   Calculate the velocity of the gases leaving the nozzle. If the nozzle isentropic efficiency is 0.97 and the mass flow rate is 90 lb/s calculate the diameter of the nozzle at exit. $c_p = 0.265$ Btu/lb R, $\gamma = 1.35$.

Answer

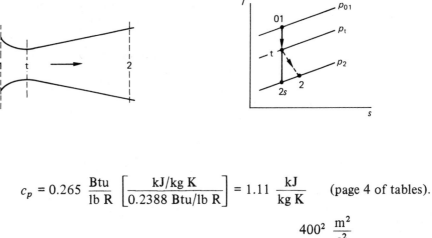

$$c_p = 0.265 \ \frac{\text{Btu}}{\text{lb R}} \left[\frac{\text{kJ/kg K}}{0.2388 \ \text{Btu/lb R}} \right] = 1.11 \ \frac{\text{kJ}}{\text{kg K}} \qquad \text{(page 4 of tables)}.$$

$$T_{01} = T_{02} = T_1 + \theta_{u1} = 880.7 \ \text{K} + \frac{400^2 \ \frac{\text{m}^2}{\text{s}^2}}{2 \times 1.11 \ \frac{\text{kJ}}{\text{kg K}} \left[\frac{\text{kg m}}{\text{N s}^2} \right] \left[\frac{10^3 \ \text{N m}}{\text{kJ}} \right]}$$

$$= 952.8 \ \text{K},$$

$$T_2 = 500 + 273 = 773 \ \text{K},$$

$$u_2 = \sqrt{2 c_p (T_{02} - T_2)} = \sqrt{2 \times 1.11 \times 179.8 \times 1000} = 631.8 \ \frac{\text{m}}{\text{s}}$$

(in the usual way for dimensions).

$$T_{01} - T_{2s} = \frac{T_{01} - T_2}{\eta} = \frac{179.8 \ \text{K}}{0.97} = 185.4 \ \text{K},$$

$$T_{2s} = 767.4 \ \text{K}.$$

$$p_{01} = p_1 \left(\frac{T_{01}}{T_1} \right)^{\gamma/(\gamma-1)} = 158.5 \ \frac{\text{kN}}{\text{m}^2} \left(\frac{952.7}{880.7} \right)^{3.857} = 214.6 \ \frac{\text{kN}}{\text{m}^2},$$

$$p_2 = p_{01} \left(\frac{T_{2s}}{T_{01}} \right)^{\gamma/(\gamma-1)} = 214.6 \ \frac{\text{kN}}{\text{m}^2} \left(\frac{767.4}{952.8} \right)^{3.857} = 93.1 \ \frac{\text{kN}}{\text{m}^2}.$$

$$v_2 = \frac{R T_2}{p_2} = \left(\frac{\gamma - 1}{\gamma} \right) c_p \left(\frac{T_2}{p_2} \right) = \frac{1}{3.857} \times 1.11 \ \frac{\text{kJ}}{\text{kg K}} \ \frac{773 \ \text{K}}{93.1 \ \frac{\text{kN}}{\text{m}^2}}$$

$$= 2.388 \ \frac{\text{m}^3}{\text{kg}}.$$

$$A_2 = \frac{\dot{m}v_2}{u_2} = \left(\frac{90}{2.205}\right)\frac{kg}{s} \times \frac{2.388 \frac{m^3}{kg}}{631.8 \frac{m}{s}} = 0.1543 \ m^2,$$

$$D_2 = \sqrt{\frac{4}{\pi} \times 0.1543 \ m^2} = 0.443 \ m.$$

Example 10.10

A stationary rocket motor consists of a reservoir, containing high-pressure gases formed by combustion of the fuel, and a suitable nozzle, through which the gases are expanded to atmospheric pressure and discharged to the atmosphere.

Under certain conditions the state of the gas in the reservoir is 2000 kN/m² and 2300 K and atmospheric pressure is 100 kN/m⁴. The area of the nozzle is 0.05 m² and the nozzle isentropic efficiency is 0.9.

Calculate the maximum mass which could be raised from the ground by this rocket operating in a gravitational field where $g = 9.80 \ m/s^2$. For the gases $m_v = 32 \ kg/kmol$ and $\gamma = 1.3$.

Answer

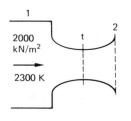

$$\frac{p_t^*}{p_{01}} = \left(\frac{2}{\gamma + 1}\right)^{\gamma/(\gamma-1)} = \left(\frac{2}{2.3}\right)^{4.333} = 0.546$$

and
$$p_t^* = 0.546 \times 2000 = 1091.5 \ \frac{kN}{m^2}.$$

Thus, since the nozzle will reach maximum mass flow when the overall pressure gradient is greater than the critical, and since the pressure in the atmosphere at 2 is greatly below the critical value at the throat, the nozzle operates under critical or choked conditions, and must be convergent–divergent.

$$R = \frac{R_0}{m_v} = \frac{8.3143 \ \frac{kJ}{kmol \ K}}{32 \ \frac{kg}{kmol}} = 0.26 \ \frac{kJ}{kg \ K}.$$

$$c_p = \left(\frac{\gamma}{\gamma - 1}\right) R = \frac{1.3 \times 0.26}{0.3} = 1.126 \ \frac{kJ}{kg \ K}.$$

$$T_{2s} = T_{01} \left(\frac{p_2}{p_{01}}\right)^{(\gamma-1)/\gamma} = 2300 \ K \left(\frac{100}{2000}\right)^{0.231} = 1151.3 \ K,$$

$$T_{01} - T_{2s} = 1148.7 \ K,$$

$$T_{01} - T_2 = \eta (T_{01} - T_{2s}) = 0.9 \times 1148.7 = 1033.8 \text{ K} \qquad (T_2 = 1266.2 \text{ K}).$$

$$u_2 = \sqrt{2c_p (T_{01} - T_2)} = \sqrt{2 \times 1.126 \frac{\text{kJ}}{\text{kg K}} (1033.8 \text{ K}) \left[\frac{\text{kg m}}{\text{N s}^2}\right]\left[\frac{10^3 \text{ N m}}{\text{kJ}}\right]} = 1526 \frac{\text{m}}{\text{s}},$$

$$v_2 = \frac{RT_2}{p_2} = \frac{0.26 \frac{\text{kJ}}{\text{kg K}} \times 1266.2 \text{ K}}{100 \frac{\text{kN}}{\text{m}}} = 3.292 \frac{\text{m}^3}{\text{kg}},$$

$$\dot{m} = \frac{u_2 A_2}{v_2} = \frac{1526 \frac{\text{m}}{\text{s}} \times 0.05 \text{ m}^2}{3.292 \frac{\text{m}^3}{\text{kg}}} = 23.18 \frac{\text{kg}}{\text{s}}.$$

Thrust $F = \dot{m}u_2$ (full expansion) $= 23.18 \frac{\text{kg}}{\text{s}} \times 1526 \frac{\text{m}}{\text{s}} \left[\frac{\text{N s}^2}{\text{kg m}}\right] = 35\,367 \text{ N}.$

Rocket mass $= \dfrac{F}{g} = \dfrac{35\,367 \text{ N}}{9.8 \frac{\text{m}}{\text{s}^2}} \left[\dfrac{\text{kg m}}{\text{N s}^2}\right] = 3609 \text{ kg}.$

Exercises

1 A rocket motor designed to give a thrust of 5000 N at an altitude of 12 200 m has a combustion chamber pressure and temperature of 21 bar and 3000 K respectively. If the pressure of the atmosphere at this altitude is 0.188 bar and γ for the combustion gases is 1.3, determine the Mach number at nozzle exit and the nozzle throat area. Assume isentropic flow throughout and take $R = 0.26$ kJ/kg K.

(*Answer:* 3.25, 1.66×10^{-3} m^3)

2 Air flows through a long pipe which is fitted with a convergent nozzle at exit. Show that, if the flow is assumed adiabatic and the velocity at the throat is sonic, then

$$T_2 = T_1 \left[\frac{\dfrac{2}{\gamma - 1} + (Ma_1)^2}{1 + \dfrac{2}{\gamma - 1}} \right],$$

where suffix 1 refers to the pipe condition and suffix 2 refers to the throat condition.

A pipe fitted with a convergent nozzle, of throat diameter 75 mm, discharges air to the atmosphere. At the throat the pressure is 100 kN/m^2 and the velocity is sonic. In the pipe the temperature is 15 °C and the Mach number is 0.4. Determine the throat temperature and the mass flow rate. $\gamma = 1.4$ and $R = 0.287$ kJ/kg K.

(*Answer:* 248 K, 1.769 kg/s)

3 The nozzles in the first stage of a steam turbine are fed from a steam chest of large cross-sectional area, in which the pressure is 110 bar and the temperature 400 °C.

Calculate:
(a) the exit velocity if the exit pressure is 40 bar and the overall nozzle isentropic efficiency is 0.96;
(b) the throat area to pass 20 kg/s of steam assuming that flow is isentropic between inlet and throat. (*Answer:* 679 m/s, 0.001 389 m^2)

4 Steam in the saturated vapour state enters a convergent nozzle with low velocity. Determine, assuming stable isentropic flow within the nozzle, the

temperature just inside the nozzle exit if the back pressure in the vessel to which the nozzle is discharging is (a) 3 bar, (b) 2 bar.

Calculate the mass discharge per unit area at exit in case (a).

(*Answer:* 133.5 °C, 128.7 °C, 653.3 kg/m² s)

5 Steam is supplied to a convergent–divergent nozzle by means of a pipe of internal diameter 120 mm. When the mass flow rate is 15 kg/s the steam conditions in the pipe, just before the nozzle entry, are 18 bar, 0.995 dry.

Calculate the pressure at the throat and the throat and exit areas required if the nozzle discharges at a pressure of 1 bar.

Assume isentropic expansion to the throat and an overall isentropic efficiency of 0.97. Take the critical pressure ratio for expansion of the two-phase mixture to be 0.579. (*Hint:* Use the *h–s* chart for ease of working).

(*Answer:* 11 bar, 0.0219 m²)

Note on Specimen Examination Papers

As already indicated in the Preface, Chapters 11 and 12 give two possible sessional examination papers covering the material included in this volume.

The content of these papers should be viewed with some discretion, since a well-balanced examination will call upon the student to present, from memory, some fundamental theoretical work in the derivation of results from first principles. The examples given here are designed to give the reader further facility in the solution of numerical problems and assume that the theoretical background has already been assimilated.

One or two problems on unsteady flow do not come within the general remit of the preceding chapters and are included to show the general method of attack. Unlike in thermodynamics, where the full unsteady-flow energy equation can be propounded and used in a logical manner with appropriate assumptions in each case, the problems on unsteady flow given herein are distinctive and do not lend themselves in any meaningful way to any general theory of unsteady-flow behaviour but have to be solved on an individual basis.

The student is strongly urged once again to attempt the problems for himself or herself before referring to the solutions. The further requirement here is of course that chapter headings are now absent, and a decision has to be taken on which particular piece of theory is under investigation and use in any given problem.

The papers comprise eight questions in each case and a typical time allowance for a completed paper (of five solutions) would be three hours.

11 Examination Paper Number 1

11.1 Questions

1 If it can be assumed that the power W required by a pump depends only on the rate of discharge $\dot{V}$, the fluid density and viscosity ρ and μ, the rotational speed N, the impeller diameter D and the difference of pressure across the pump Δp, show, by dimensional analysis, that $\dot{W}$ may be expressed as:

$$\dot{W} = \rho N^3 D^5 \phi \left(\frac{\rho N D^2}{\mu} , \frac{\Delta p}{\rho N^2 D^2} , \frac{\dot{V}}{N D^3} \right).$$

The following results were obtained from a test on pump A, discharging water:

$N = 1000 \text{ rev/min}; \Delta p = 120 \text{ kN/m}^2 ; \dot{V} = 0.015 \text{ m}^3/\text{s}; \dot{W} = 2.2 \text{ kW}.$

Calculate the power required, the pressure rise and the speed of a second pump B, whose impeller diameter is twice that of A, discharging water at the rate of 0.045 m³/s under dynamically similar conditions. Ignore the effect of Reynolds number.

2 A cylindrical tank, 2 m in diameter and 4 m high, has an orifice of diameter 150 mm in the base which has a coefficient of discharge of 0.60. Water is continually flowing into the tank at the rate of 20 litre/s. Find the time taken for the level of the water to fall from 1.5 to 0.5 m.

3 Draw a graph of friction factor f against Reynolds number (Re) (log scale) for flow through pipes of various roughness. Indicate the regions of laminar, smooth turbulent and fully rough turbulent flow.

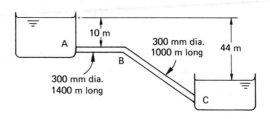

A pipeline ABC connects two reservoirs as shown in the diagram. If the friction factor f is 0.008 determine the water flow rate and the gauge pressure at B for the conditions shown. Neglect all energy degradation except that due to pipe friction.

4 A venturimeter is calibrated by passing water through it and observing the readings of a mercury manometer connected between the throat and entry sections which are 914 mm apart. When 15 l/min are passing through the venturimeter the gauge reading is 356 mm.

Subsequently the meter is installed in a vertical pipe to measure the flow rate of petrol in an upward direction. Two Bourdon pressure gauges are used in place of the manometer and the difference in their gauge readings is observed to be 20.7 kN/m².

Calculate the flow rate through the meter, ignoring friction and working from the Bernoulli equation. The specific gravities of petrol and mercury are 0.78 and 13.6 respectively.

5 The diagram shows a U-tube containing mercury which is used in conjunction with a venturimeter to measure the flow of water along a horizontal pipe. End C is connected to the inlet of the venturimeter and end B to the throat. A is a cylindrical vessel with a cross-sectional area 40 times that of the tube.

The law of the venturimeter is

$$\dot{V} = 0.003H^{1/2},$$

where $\dot{V}$ is the volumetric flow rate in m³/s and H is the difference of pressure head in m water between the inlet and the throat.

Calculate the distance y through which the surface of separation will move from its zero position when $\dot{V} = 0.006$ m³/s. The specific gravity of mercury is 13.6.

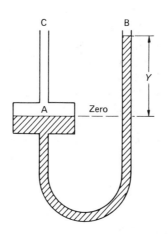

6 Oil flows down an inclined plane under laminar flow conditions. If the film thickness, t, is maintained constant, show by considering a small element of oil (see diagram) that

$$\mu \frac{d^2 u}{dy^2} = -\rho g \sin \theta.$$

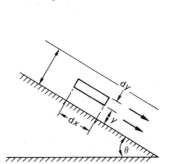

Hence, by integration, show that if the gradient of the plane is 30° and the oil film thickness is 5 mm then the maximum velocity will be 1.53 m/s and the oil flow rate per metre width of plane is 0.005 m³/s. For the oil, the kinematic viscosity, ν, is 0.000 04 m²/s. It may be assumed that the air resistance at the oil surface is negligible.

7 A jet of water is directed at a uniform flat rectangular plate, of mass 5 kg, which is suspended from a hinge along the top horizontal edge (see diagram). If the water nozzle has a coefficient of discharge of 0.97, determine the pressure, p_0, required in the pipe, upstream of the nozzle, to maintain equilibrium with the plate inclined at 30° to the vertical.

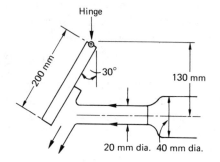

8 An inverted U-tube acts as a siphon. It has one end immersed in water in a tank, and the other end is free to discharge to atmosphere. The free end is 3.5 m beneath the top end of the siphon, which is 3 m above the free surface of the water in the tank.

   If the tube forming the siphon is of uniform bore and energy degradation due to friction is negligible, determine:
   (a) the velocity of the water,
   (b) the pressure at the top of the siphon.

11.2 Suggested Solutions

Question 1

$$\dot{W} = \phi(\dot{V}, \rho, \mu, N, D, \Delta p) = V^a \rho^b \mu^c N^d D^e \Delta p^f + \text{etc.}$$

Thus $\dfrac{ML^2}{t^2} = \left(\dfrac{L^3}{t}\right)^a \left(\dfrac{L^3}{M}\right)^b \left(\dfrac{M}{Lt}\right)^c \left(\dfrac{1}{t}\right)^d L^e \left(\dfrac{M}{Lt^2}\right)^f.$

Mass $1 = c + f - b;$

Length $2 = 3a + 3b - c + e - f.$ Get d, e and f in terms of a, b and c.

Time $-3 = -a - c - d - 2f.$

$f = b - c + 1,$

$e = 2 - 3a - 3b + c + f = 2 - 3a - 3b + c + b - c + 1 = 3 - 3a - 2b;$

$d = 3 - a - c - 2f = 3 - a - c - 2(b - c + 1) = 1 - a - 2b + c.$

$\dot{W} = \dot{V}^a \rho^b \mu^c N^{1-a-2b+c} D^{3-3a-2b} \Delta p^{b-c+1}$

$= \rho N^3 D^5 [\dot{V}^a \rho^{b-1} N^{-2-a-2b+c} \mu^c D^{-2-3a-2b} \Delta p^{b-c+1}]$

$= \rho N^3 D^5 \left[\left(\dfrac{\dot{V}}{ND^3}\right)^a \left(\dfrac{\rho ND^2}{\mu}\right)^{-c} \rho^{-c+b-1} N^{-2-2b+2c} D^{-2-2b+2c} \Delta p^{b-c+1}\right]$

$$= \rho N^3 D^5 \left[\left(\frac{\dot{V}}{ND^3} \right)^a \left(\frac{\rho ND^2}{\mu} \right)^{-c} \left(\frac{\Delta p}{\rho N^2 D^2} \right)^{b-c+1} \right]$$

$$= \rho N^3 D^5 \phi \left[\left(\frac{\rho ND^2}{\mu} \right), \left(\frac{\Delta p}{\rho N^2 D^2} \right), \left(\frac{\dot{V}}{ND^3} \right) \right].$$

$(Re) = \dfrac{\rho ND^2}{\mu}$ is constant.

$$\left(\frac{\dot{V}}{ND^3} \right)_A = \left(\frac{\dot{V}}{ND^3} \right)_B \quad \text{or} \quad N_B = N_A \left(\frac{D_A}{D_B} \right)^3 \left(\frac{\dot{V}_B}{\dot{V}_A} \right) = 1000 \, \frac{\text{rev}}{\text{min}} \times \frac{1}{8} \times 3 = 375 \, \frac{\text{rev}}{\text{min}}.$$

$$\left(\frac{\Delta p}{\rho N^2 D^2} \right)_A = \left(\frac{\Delta p}{\rho N^2 D^2} \right)_B \quad \text{or} \quad \Delta p_B = \Delta p_A \left(\frac{N_B}{N_A} \right)^2 \left(\frac{D_B}{D_A} \right)^2 = 120 \, \frac{\text{kN}}{\text{m}^2} \times 0.375^2 \times 4 = 67.5 \, \frac{\text{kN}}{\text{m}^2}.$$

$$\left(\frac{\dot{W}}{\rho N^3 D^5} \right)_A = \left(\frac{\dot{W}}{\rho N^3 D^5} \right)_B \quad \text{or} \quad \dot{W}_B = \dot{W}_A \left(\frac{N_B}{N_A} \right)^3 \left(\frac{D_B}{D_A} \right)^5 = 2.2 \, \text{kW} \times 0.375^3 \times 32 = 3.71 \, \text{kW}.$$

Question 2

Volume change in vessel in time $dt = (\dot{V}_{in} - \dot{V}_{out}) \, dt$

Volume change in vessel in time $dt = A \, dh$

where A = vessel cross-sectional area.

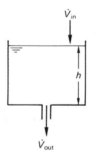

Thus $\qquad (0.02 - aC_D \sqrt{2gh}) \, dt = \frac{1}{4} \pi \times 4 \, dh$

where a = orifice area = $\frac{1}{4} \pi \times 0.0225 \, \text{m}^2$,

Thus $dt = \dfrac{\pi dh}{(0.02 - \frac{1}{4} \pi \times 0.0225 \times 0.6 \sqrt{2g} \sqrt{h})}$,

$$t = \left[\int_{1.5}^{0.5} \frac{\pi \, dh}{0.02 - \frac{1}{4} \times 0.0135 \sqrt{2g} \sqrt{h}} \right] = \left[\int \frac{\pi \, dh}{0.02 - 0.047 \sqrt{h}} \right]_{1.5}^{0.5}$$

Solution by Simpson's rule (approximate)

h	=	1.5	1.0	0.5
$\sqrt{h}$	=	1.2247	1.0	0.7071
$0.047\sqrt{h}$	=	0.0576	0.047	0.0332
$0.02 - 0.047\sqrt{h}$	=	-0.0376	-0.027	-0.0132
$\pi/(0.02 - 0.047\sqrt{h})$	=	-83.56	-116.370	-238.0

$t = \dfrac{4h}{3} (\phi_1 + 4\phi_2 + \phi_3) = \dfrac{-0.5}{3} [-83.56 - (4 \times 116.37) - 238] = 131.25$ approx.

OR, by integration, $\displaystyle\int_{h=1.5}^{h=0.5} \frac{dx}{a - b\sqrt{x}} = \frac{-2}{6}\left[\sqrt{x} + \frac{a}{b} \ln(a - b\sqrt{x})\right]$,

$$t = \pi\left[\frac{2}{0.047} \left(\sqrt{1.5} - \sqrt{0.5}\right) + \frac{2 \times 0.02}{0.047^2} \ln \frac{0.02 - 0.047\sqrt{1.5}}{0.02 - 0.047\sqrt{0.5}}\right]$$

$$= \pi\left[22 + 18.12 \ln \frac{0.0375}{0.0132}\right] = 40.8\pi = 128.2 \text{ s exactly.}$$

Question 3

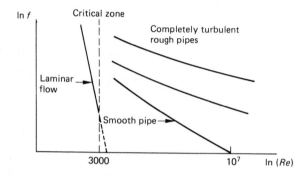

Applying Bernoulli from 1 and 2 gives

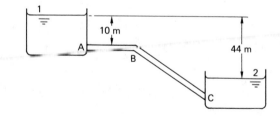

$$44 \text{ m} = 4 \frac{fL}{D} \frac{u^2}{2g},$$

or

$$\frac{u^2}{2g} = \frac{44 \text{ m} \times 0.3 \text{ m}}{4 \times 0.008 \times 2400 \text{ m}} = \frac{11}{64} \text{ m};$$

$$u = \sqrt{2 \times 9.81 \frac{\text{m}}{\text{s}^2} \times \frac{11}{64} \text{ m}} = 1.88 \frac{\text{m}}{\text{s}}.$$

$$\dot{V} = Au = \tfrac{1}{4}\pi \times 0.09 \text{ m}^3 \times 1.88 \frac{\text{m}}{\text{s}} = 0.133 \frac{\text{m}^3}{\text{s}}.$$

Applying Bernoulli between 1 and B gives

$$10 \text{ m} = \frac{p_B}{\rho g} + \frac{u^2}{2g} \left(\frac{4fL_{AB}}{D} \right);$$

$$\frac{p_B}{\rho g} = 10 - \frac{11}{64} \left(1 + \frac{4 \times 0.008 \times 1400}{0.3} \right) = -15.9 \text{ m},$$

$$p_B = -10^3 \frac{\text{kg}}{\text{m}^3} \times 9.81 \frac{\text{m}}{\text{s}^2} \times 15.9 \text{ m} \left[\frac{\text{N s}^2}{\text{kg m}} \right] = -156 \frac{\text{kN}}{\text{m}^2}.$$

Question 4

For a venturimeter, $\qquad \dot{V} = \dfrac{A_2 C_D \sqrt{2g h_m}}{\sqrt{1 - \left(\dfrac{A_2}{A_1} \right)^2}} = K \sqrt{h_m},$

when $\dot{V} = 15 \times 10^{-3} \dfrac{\text{m}^3}{\text{min}}$, gauge reading = 0.356 m.

Thus $\qquad h_m = h \left(\dfrac{\rho_m}{\rho} - 1 \right) = 12.6 \times 0.356 \text{ m},$

$$K = \frac{15 \times 10^{-3} \dfrac{\text{m}^3}{\text{min}} \left[\dfrac{\text{min}}{60 \text{ s}} \right]}{\sqrt{12.6 \times 0.356} \text{ m}^{1/2}} = 0.118 \times 10^{-3} \frac{\text{m}^{5/2}}{\text{s}}.$$

When meter is mounted vertically,

$$h_m = \frac{p_1 - p_2}{\rho g} + z_1 - z_2 = \frac{20.7 \dfrac{\text{kN}}{\text{m}^2} \times \text{s}^2}{0.78 \times 10^3 \dfrac{\text{kg}}{\text{m}^3} \times 9.81 \text{ m}} \left[\frac{\text{kg m}}{\text{N s}^2} \right] - 0.914 \text{ m}$$

$$= 1.791 \text{ m}.$$

$$\dot{V} = K \sqrt{h_m} = 0.118 \times 10^{-3} \sqrt{1.791} \frac{\text{m}^3}{\text{s}} = 0.158 \times 10^{-3} \frac{\text{m}^3}{\text{s}}.$$

Question 5

Venturimeter $\dot{V} = 0.006 \dfrac{\text{m}^3}{\text{s}} = 0.003 H^{1/2}.$

Thus $H = 4$ m water.

$$(p_2 - p_1) = \rho g H = 10^3 \frac{\text{kg}}{\text{m}^3} \times 9.81 \frac{\text{m}}{\text{s}^2} \times 4 \text{ m} \left[\frac{\text{N s}^2}{\text{kg m}} \right]$$

$$= 39\,240 \frac{\text{N}}{\text{m}^2}.$$

For the manometer pressure along 1–1 constant,

$$p_2 + \rho g (z + x) = p_1 + (z - y) \rho g + \rho_m g (y + x),$$

$$p_2 - p_1 = -\rho g (x + y) + \rho_m g (x + y) = (x + y) g (\rho_m - \rho).$$

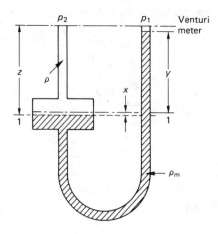

Also from continuity, $\qquad Ax = ay,$

or $\qquad\qquad\qquad\qquad x = \dfrac{a}{A} y = \dfrac{y}{40}.$

$$p_2 - p_1 = 39\,240 \ \frac{N}{m^2} = y \times \frac{41}{40} g (13.6 - 1) \ 10^3 \ \frac{kg}{m^3}.$$

$$y = 4 \times \frac{40}{41} \times \frac{1}{12.6} = 0.31 \ m \qquad \left(\text{since } \frac{p_2 - p_1}{\rho g} = 4 \ m \right).$$

Question 6

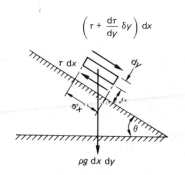

Resolving along the plane, for equilibrium, gives

$$\left(\tau + \frac{d\tau}{dy} \ dy \right) dx - \tau \ dx + \rho g \ dx \ dy \sin \theta = 0$$

or $\qquad\qquad \dfrac{d\tau}{dy} \ dy \ dx + \rho g \ dx \ dy \sin \theta = 0$

or $\qquad\qquad \dfrac{d\tau}{dy} + \rho g \sin \theta = 0.$

But $\qquad\qquad\qquad \tau = \mu \ \dfrac{du}{dy} \qquad\qquad$ (Newton).

Thus
$$\mu \frac{d^2u}{dy^2} + \rho g \sin \theta = 0 \qquad \text{(for constant } \mu)$$

$$\mu \frac{d^2u}{dy^2} = -\rho g \sin \theta.$$

Integrating, we get
$$\mu \frac{du}{dy} + \rho gy \sin \theta = A$$

and
$$\mu u + \rho g \frac{y^2}{2} \sin \theta = Ay + B.$$

For $y = 0, u = 0$;   thus $B = 0$.

For $y = t, \frac{du}{dy} = 0$;   thus $A = \rho g t \sin \theta$.

Thus
$$\mu u + \rho g \frac{y^2}{2} \sin \theta = \rho g t y \sin \theta$$

and
$$u = \frac{\rho g \sin \theta \, (2ty - y^2)}{2\mu}.$$

Now $d\dot{V} = u\,dy$

and thus
$$\dot{V} = \int_0^t u\,dy = \int_0^t \frac{\rho g \sin \theta \, (2ty - y^2)}{2\mu} = \frac{\rho g \sin \theta}{2\mu} \left[ty^2 - \frac{y^3}{2} \right]_0^t$$

$$= \frac{\rho g \sin \theta \, (t^3)}{3\mu}.$$

$\hat{u}$ at surface, $y = t$;

$$\hat{u} = \frac{\rho g \sin \theta \, (t^2)}{2\mu} = \frac{9.81 \, \frac{m}{s^2} \times 0.5 \times (0.005)^2 \, m^2}{2 \times 0.000\,04 \, \frac{m^2}{s}} = 1.533 \, \frac{m}{s}.$$

$$\dot{V} = \frac{9.81 \, \frac{m}{s^2} \times 0.5 \, (0.005)^3 \, m^3}{3 \times 0.000\,04 \, \frac{m^2}{s}} = 0.005\,11 \, \frac{m^3}{s} \text{ per metre width of plane.}$$

Question 7

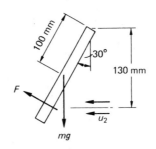

Taking moments about the hinge gives

$$F \frac{130 \text{ mm}}{\cos 30°} = mg \, 100 \text{ mm} \times \sin 30°,$$

$$F = 5 \text{ kg} \times 9.81 \ \frac{\text{m}}{\text{s}^2} \times \frac{100}{130} \sin 30° \cos 30° \left[\frac{\text{N s}^2}{\text{kg m}} \right] = 16.34 \text{ N}.$$

Momentum equation: $F = \dot{m}u_2 \cos 30° = \rho A u_2^2 \cos 30°.$

$$u_2 = \sqrt{\frac{F}{\rho A \cos 30°}} = \sqrt{\frac{16.34 \text{ N} \times 4}{10^3 \ \frac{\text{kg}}{\text{m}^3} \times \pi \times 0.02^2 \text{ m}^2 \times 0.866} \left[\frac{\text{kg m}}{\text{N s}^2} \right]}$$

$$= 7.75 \ \frac{\text{m}}{\text{s}}.$$

For an ideal fluid, Bernoulli's equation gives

$$\frac{p_1}{\rho g} + \frac{u_1^2}{2g} = \frac{u_2^2}{2g}.$$

Continuity: $\qquad\qquad\qquad A_1 u_1 = A_2 u_2$

or $\qquad\qquad\qquad\qquad u_2 = 4u_1.$

Thus $\qquad\qquad\qquad \dfrac{p_1}{\rho g} = \dfrac{u_2^2 - u_1^2}{2g} = \dfrac{15}{16} \dfrac{u_2^2}{2g}.$

Thus for real fluid ($C_D = 0.97$),

$$p_1 = \rho \ \frac{15}{16} \ \frac{u_2^2}{2} \ \frac{1}{0.97^2} = 10^3 \ \frac{\text{kg}}{\text{m}^3} \ \frac{15}{16} \times \frac{7.75^2 \text{ m}^2}{2 \times 0.97^2} \left[\frac{\text{N s}^2}{\text{kg m}} \right]$$

$$= 29.92 \ \frac{\text{kN}}{\text{m}^2}.$$

Question 8

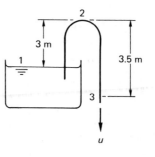

Applying Bernoulli's equation from 1 to 3 gives

$$\frac{p_1}{\rho g} + \frac{u_1^2}{2g} + z_1 = \frac{p_3}{\rho g} + \frac{u_3^2}{2g} + z_3,$$

i.e. $\qquad\qquad\qquad z_1 - z_3 = \dfrac{u_3^2}{2g} \qquad$ (all else is zero).

$$u_3 = \sqrt{2 \times g(z_1 - z_3)} = \sqrt{2 \times 0.5 \text{ m} \times 9.81 \ \frac{\text{m}}{\text{s}^2}}$$

$$= 3.13 \ \frac{\text{m}}{\text{s}}.$$

Applying Bernoulli's equation from 2 to 3 gives

$$\frac{p_2}{\rho g} + \frac{u_2^2}{2g} + z_2 = \frac{p_3}{\rho g} + \frac{u_3^2}{2g} + z_3 \qquad (p_3 = p_{atm})$$

$$u_2 = u_3 \qquad \text{(continuity)}.$$

$$\frac{p_2 - p_{atm}}{\rho g} = z_3 - z_2 = -3.5 \text{ m}$$

$$p_2 - p_{atm} = -3.5 \text{ m} \times 10^3 \frac{\text{kg}}{\text{m}^3} \times 9.81 \frac{\text{m}}{\text{s}^2} \left[\frac{\text{N s}^2}{\text{kg m}}\right] = -34.34 \frac{\text{kN}}{\text{m}^2} .$$

12 Examination Paper Number 2

12.1 Questions

1 Show, by dimensional analysis, that the torque τ required to rotate a disc of diameter D at an angular velocity ω, in a fluid of density ρ and viscosity μ, is given by:

$$\tau = \rho D^5 \omega^2 \phi \left(\frac{\rho D^2 \omega}{\mu}\right).$$

A disc A of diameter 0.30 m requires a torque of 10 N m when rotating in water at 300 rad/s. Another disc B, of diameter 0.90 m, is to be tested in air under dynamically similar conditions. Calculate:
(a) the corresponding speed of disc B in rev/min,
(b) the torque required by disc B at that speed,
(c) the power required to drive disc B.

For water: $\rho = 1\,000$ kg/m^3 and $\mu = 0.001\,013$ N s/m^2.
For air:　　$\rho = 1.25$ kg/m^3 and $\mu = 0.000\,018\,5$ N s/m^2.

2 A pump draws water from a sump and discharges it into a tank in which the level of water is 80 m above that in the sump (see diagram). The internal diameters of the suction and discharge pipes are 100 mm and 50 mm respectively. The inlet and exit sections of the pump are in the same horizontal plane 6 m above the water level in the sump.
　　The energy degradation is as follows:

In the suction pipe it is twice the velocity head in that pipe.
In the discharge pipe it is 25 times the velocity head in that pipe.
At suction pipe inlet it is 0.5 times the velocity head in that pipe.
At discharge pipe exit it is equal to the velocity head in that pipe.

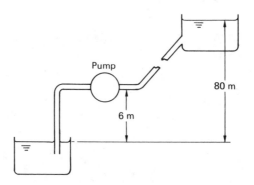

172

When the power transmitted to the water by the pump is 40 kW, the pressure at inlet to the pump is ($-$ 7 m) water gauge. For these conditions calculate the flow rate through the pump and the energy degradation due to friction between the pump inlet and outlet sections.

3 A vertical cylindrical tank containing water discharges through an orifice in its base. Derive an expression for the time taken to lower the head above the orifice from H_1 to H_2.

A hemispherical cistern is 4 m in diameter and full of water. How long will it take for the contents to be discharged through a sharp-edged orifice, of diameter 80 mm, in the bottom of the cistern? Take the coefficient of discharge to be 0.6.

4 A vertical pipeline gradually narrows from a diameter of 0.6 m to one of 0.3 m while rising through a distance of 1.5 m.

Calculate the magnitude and direction of the vertical force on this pipe contraction when a flow of 0.3 m^3/s of water passes upwards through the pipeline and the pressure at the larger diameter is 200 kN/m^2 gauge. Assume that energy degradation due to friction in the contraction is equal to 0.5 ($u_1^2/2g$), where u_1 is the inlet flow velocity.

5 Two concentric cylinders are rotated in opposite directions, each at a speed of 100 rev/min. The inner cylinder is of diameter 100 mm and the clearance between the cylinders is 1 mm. If the viscosity of oil within the clearance space is 0.15 N s/m^2, calculate the torques required to rotate the cylinders if they are both 50 mm long. State any assumptions made.

If the clearance between the cylinders had been much larger (say 50 mm) discuss how much this would have affected the theory.

6 A reservoir containing water, oil and air under pressure is closed by a rectangular gate AB, 2 m long and 1 m wide, hinged at A (see diagram). If the gate is of negligible weight, determine:
(a) the magnitude and point of application of the resultant force on the gate,
(b) the magnitude of the vertical force F to be applied at B to hold the gate in the closed position.

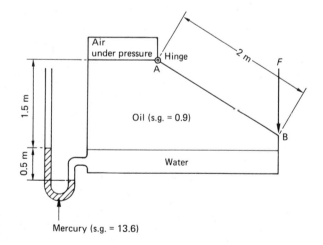

7 Oil, having a specific gravity of 0.9, and viscosity 0.03 N s/m^2, is pumped through a pipe of diameter 0.25 m. The velocity of the oil at the pipe centre line is determined using Pitot and static pressure measurements, the pressure tappings being connected to a U-tube containing mercury.

If the difference of levels in the limbs of the manometer is 100 mm, determine the velocity of the oil.

Take the coefficient of the Pitot tube as unity and the specific gravity of mercury as 13.6.

If the ratio of u(mean)/u(centre) is 0.50 for laminar flow and 0.80 for turbulent flow, obtain the mass flow rate of oil through the pipe.

8 Water flows along a V-shaped open channel under uniform steady conditions to a depth of 1 m. The sides of the channel are inclined to the horizontal at an angle of 60°. Calculate the rate of flow along the channel if its slope is 1 in 2500 and the value of the Chezy constant is 60 $m^{1/2}$/s.

After a period of time the channel becomes silted up to a depth equal to that of the water above it. What is now the depth of water above the silt if both the rate of flow along the channel and the Chezy constant remain unaltered?

12.2   Suggested Solutions

Question 1

$$\tau = \phi\,(D, \omega, \rho, \mu) = D^a \omega^b \rho^c \mu^d + \text{etc.}$$

$$\frac{ML^2}{t^2} = L^d\,\frac{1}{t^b}\,\frac{M^c}{L^{3c}}\,\frac{M^d}{L^d t^d}\,.$$

Mass	$1 = c + d$	or		$c = 1 - d.$
Length	$2 = a - 3c - d$		or	$a = 2 + 3c + d = 2 + 3 - 3d + d = 5 - 2d.$
Time	$-2 = -b - d$		or	$b = 2 - d.$

$$\tau = D^{5-2d} \omega^{2-d} \rho^{1-d} \mu^d$$

$$= \rho D^5 \omega^2\,[D^{-2d} \omega^{-d} \rho^{-d} \mu^d]$$

$$= \rho D^5 \omega^2\,\left[\left(\frac{\rho D^2 \omega}{\mu}\right)^{-d}\right]$$

$$= \rho D^5 \omega^2 \phi\left(\frac{\rho D^2 \omega}{\mu}\right).$$

$$\left(\frac{\rho D^2 \omega}{\mu}\right)_{\text{water}} = \left(\frac{\rho D^2 \omega}{\mu}\right)_{\text{air}}$$

or $\quad \omega_a = \omega_w \left(\frac{\rho_w}{\rho_a}\right)\left(\frac{D_w}{D_a}\right)^2\left(\frac{\mu_a}{\mu_w}\right) = 300\,\frac{\text{rad}}{\text{s}} \times \left(\frac{1000}{1.25}\right)\left(\frac{1}{3}\right)^2\left(\frac{0.185}{10.13}\right)$

$\qquad = 487\,\dfrac{\text{rad}}{\text{s}}\,.$

$$N_a = 487\,\frac{\text{rad}}{\text{s}}\left[\frac{\text{rev}}{2\pi\,\text{rad}}\right]\left[\frac{60\,\text{s}}{\text{min}}\right] = 4650\,\frac{\text{rev}}{\text{min}}\,.$$

$$\left(\frac{\tau}{\rho D^5 \omega^2}\right)_a = \left(\frac{\tau}{\rho D^5 \omega^2}\right)_w$$

$\tau_a = \tau_w\left(\dfrac{\rho_a}{\rho_w}\right)$ or $\left(\dfrac{D_a}{D_w}\right)^5\left(\dfrac{\omega_a}{\omega_w}\right)^2 = 10\,\text{N m}\left(\dfrac{1.25}{1000}\right)(3)^5\left(\dfrac{487}{300}\right)^2 = 8\,\text{N m}.$

$$\text{Power }\dot{W}_B = \tau_B \omega_B = 8\,\text{N m} \times 487\,\frac{\text{rad}}{\text{s}}\left[\frac{\text{W s}}{\text{N m}}\right]\left[\frac{\text{kW}}{10^3\,\text{w}}\right] = 3.9\,\text{kw.}$$

Question 2

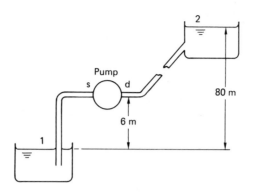

Applying Bernoulli's equation between 1 and s in the suction pipe gives

$$0 = 6 + \frac{u_s^2}{2g} + \frac{p_s}{\rho g} + h_{f,1\text{-}s}.$$

For $\dot{W} = 40$ kW, $\dfrac{p_s}{\rho g} = -7$ m and $h_{f,1\text{-}s} = (2 + 0.5)\dfrac{u_s^2}{2g}$.

Thus $(7 - 6)$ m $= 1$ m $= 3.5\dfrac{u_s^2}{2g}$,

$$u_s = \sqrt{\frac{2g}{3.5}} = 2.3676\ \frac{\text{m}}{\text{s}}.$$

$$\dot{V}_s = A_s u_s = \frac{\pi}{4} \times 0.1^2\ \text{m}^2 \times 2.3676\ \frac{\text{m}}{\text{s}} = 0.0186\ \frac{\text{m}^3}{\text{s}}.$$

Now $\dot{W} = 40$ kW $= \rho \dot{V} g H_T$.

Thus
$$H_T = \frac{40 \times 10^3\ \text{W}}{10^3\ \dfrac{\text{kg}}{\text{m}^3} \times 0.0186\ \dfrac{\text{m}^3}{\text{s}} \times 9.81\ \dfrac{\text{m}}{\text{s}^2}} \left[\frac{\text{N m}}{\text{W s}}\right]\left[\frac{\text{kg m}}{\text{N s}^2}\right]$$

$$- 219.3\ \text{m}.$$

Applying Bernoulli between 1 and 2 gives

$$219.3\ \text{m} = 80\ \text{m} + 2.5\ \frac{u_s^2}{2g} + \frac{26u_d^2}{2g} + h_{f,s\text{-}d}.$$

$$u_d = u_s \frac{A_s}{A_d} = u_s \left(\frac{D_s}{D_d}\right)^2 = 2.3676\ \frac{\text{m}}{\text{s}} \times 4 = 9.4704\ \frac{\text{m}}{\text{s}}.$$

$$h_{f,s\text{-}d} = 219.3 - 80 - 2.5\left(\frac{2.3676^2}{2 \times 9.81}\right) - 26\left(\frac{9.4704^2}{2 \times 9.81}\right)$$

$$= 139.3 - 0.714 - 118.86$$

$$= 19.73\ \text{m}.$$

Question 3

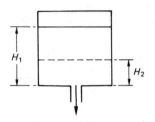

A = cylinder cross-sectional area, $\qquad$ a = orifice cross-sectional area.

$$-\dot{V} = C_{\mathrm{D}}a\sqrt{2gH}$$

at a given head H through the orifice.

Thus $\qquad\qquad\qquad\qquad -\dfrac{\mathrm{d}\dot{V}}{\mathrm{d}t} = C_{\mathrm{D}}a\sqrt{2gH}$

or $\qquad\qquad -\displaystyle\int_{H_1}^{H_2} \dfrac{\mathrm{d}\,(HA)}{C_{\mathrm{D}}a\sqrt{2gH}} = \int_0^t \mathrm{d}t \qquad$ (but A = constant).

Thus $\qquad \displaystyle\int_0^t \mathrm{d}t = \int_{H_2}^{H_1} \dfrac{A}{C_{\mathrm{D}}a\sqrt{2g}}\,\dfrac{\mathrm{d}H}{H^{1/2}} \qquad$ (changing sign and limits).

Thus $\qquad t = \dfrac{A}{C_{\mathrm{D}}a\sqrt{2g}}\left(\dfrac{H^{1/2}}{\frac{1}{2}}\right)_{H_2}^{H_1} = \dfrac{2A}{C_{\mathrm{D}}a\sqrt{2g}}\,(H_1^{1/2} - H_2^{1/2}).$

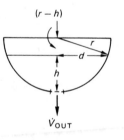

a = orifice cross-sectional area, $\qquad$ A = surface area of water at head h.

$$-\dot{V}_{\mathrm{OUT}}\,\mathrm{d}t = A\,\mathrm{d}h.$$

or $\qquad\qquad\qquad\qquad -aC_{\mathrm{D}}\sqrt{2gh} = \pi d^2\,\mathrm{d}h.$

$$d^2 = r^2 - (r - h)^2 = 2rh - h^2,$$

$$\mathrm{d}t = \dfrac{-\pi\,(2rh - h^2)\,\mathrm{d}h}{aC_{\mathrm{D}}\sqrt{2gh^{1/2}}} = \dfrac{-\pi}{aC_{\mathrm{D}}\sqrt{2g}}\,(2rh^{1/2} - h^{3/2})\,\mathrm{d}h,$$

$$t = \dfrac{-\pi}{aC_{\mathrm{D}}\sqrt{2g}}\left[\dfrac{4}{3}rh^{3/2} - \dfrac{2}{5}h^{5/2}\right]_r^0 = \dfrac{14\pi r^{5/2}}{15aC_{\mathrm{D}}\sqrt{2g}}$$

$$= \dfrac{14\pi \times 2^{5/2}\ \mathrm{m}^{5/2} \times 4}{15 \times \pi \times 0.08^2\ \mathrm{m}^2 \times 0.6\sqrt{2 \times 9.81\ \dfrac{\mathrm{m}}{\mathrm{s}^2}}} = \dfrac{995.2\ \mathrm{s}}{0.802} = 1242\ \mathrm{s}\ (20.7\ \mathrm{min}).$$

Question 4

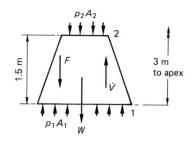

$$u_1 = \frac{\dot{V}}{A_1} = \frac{0.3 \; \frac{m^3}{s}}{\frac{1}{4} \pi \times 0.36 \; m^2} = 1.06 \; \frac{m}{s}, \qquad u_2 = 4u_1 = 4.24 \; \frac{m}{s}.$$

Applying Bernoulli's equation between 1 and 2,

$$\frac{p_1}{\rho g} = \frac{u_1^2}{2g} + z_1 = \frac{p_2}{\rho g} + \frac{u_2^2}{2g} + z_2 + h_f,$$

$$\frac{p_2}{\rho g} = \frac{p_1}{\rho g} + \frac{u_1^2}{2g} - \frac{u_2^2}{2g} + (z_1 - z_2) - h_f$$

$$= \frac{200 \; \frac{kN}{m^2}}{10^3 \; \frac{kg}{m^3} \times 9.81 \; \frac{m}{s^2}} \left[\frac{kg \; m}{N \; s^2} \right] + \frac{(1.06^2 - 4.24^2) \; \frac{m^2}{s^2}}{2 \times 9.81 \; \frac{m}{s^2}} - 1.5 \; m - \frac{0.5 \times 1.06^2 \; \frac{m^2}{s^2}}{2 \times 9.81 \; \frac{m^2}{s^2}}$$

$$= (20.4 - 0.859 - 1.5 - 0.0286) \; m = 18.01 \; m.$$

$$p_2 = 10^3 \; \frac{kg}{m^3} \times 9.81 \; \frac{m}{s^2} \times 18.01 \; m \left[\frac{N \; s^2}{kg \; m} \right] = 176\,700 \; \frac{N}{m^2}.$$

$$W = mg \;(\text{for water}) = \tfrac{1}{3} \pi \, (r_1^2 h_1 - r_2^2 h_2) \, \rho g$$

$$= \tfrac{1}{3} \pi \, (0.3^2 \; 3 - 0.15^2 \times 1.5) \; m^3 \times 9.81 \; \frac{m}{s^2} \times 10^3 \; \frac{kg}{m^3} \left[\frac{N \; s^2}{kg \; m} \right]$$

$$= 2427 \; N.$$

Momentum equation: $\quad p_1 A_1 - F - p_2 A_2 - W = \rho \dot{V} (u_2 - u_1).$

$$F = -\rho \dot{V} (u_2 - u_1) + p_1 A_1 - p_2 A_2 - W$$

$$= -10^3 \; \frac{kg}{m^3} \times 0.3 \; \frac{m^3}{s} \, (4.24 - 1.06) \; \frac{m}{s} \left[\frac{N \; s^2}{kg \; m} \right] + 200\,000 \; \frac{N}{m^2} \left(\frac{\pi}{4} \times 0.6^2 \right) m^2$$

$$-176\,700 \; \frac{N}{m^2} \left(\frac{\pi}{4} \times 0.3^2 \right) m^2 - 2427 \; N$$

$$= (-954 + 56\,549 - 12\,490 - 2427) \; N$$

$$= 40\,678 \; N \quad \text{downwards on fluid (upwards on contraction).}$$

Question 5

Shear stress $\tau = \mu \dfrac{du}{dy} = \mu \dfrac{r\omega}{t}$ (assuming parallel plate theory holds true)

where r = radius at which τ acts,
       t = thickness of clearance space, and
       ω = angular velocity (relative) = 200 rev/min.

Thus   shear force $F = \tau A$ (A = area over which τ is acting)

and   torque $= Fr = \tau A r = \mu \dfrac{r^2 \omega}{t} 2\pi r L = 2\pi \mu r^3 L \dfrac{\omega}{t}$.

Thus for inner cylinder,

$$\text{torque} = \frac{2\pi \times 0.15\ \dfrac{\text{N s}}{\text{m}^2} \times 0.1^3\ \text{m}^3 \times 0.05\ \text{m} \times 200\ \dfrac{\text{rev}}{\text{min}} \left[\dfrac{2\pi\ \text{rad}}{\text{rev}}\right] \left[\dfrac{\text{min}}{60\ \text{s}}\right]}{0.001\ \text{m}}$$

      $= 0.987$ N m.

For outer cylinder,   torque $= \dfrac{2\pi \times 0.15 \times 0.101^3 \times 0.05 \times 200 \times 2\pi}{0.001 \times 60}$

         $= 1.017$ N m.

(*Note:* The torques should be equal and opposite.)

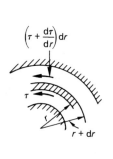

For a large clearance, equating the torques on a small annular element (shaded in the diagram),

$$\tau r^2 = \left(\tau + \frac{d\tau}{dr}\ dr\right)(r + dr)^2$$

or $0 = 2r\tau\ dr + \dfrac{d\tau}{dr}\ r^2\ dr$

or $-r\ d\tau = 2\tau\ dr.$

Thus $-\displaystyle\int \frac{d\tau}{\tau} = \int 2\ \frac{dr}{r}$

                  $-\ln \tau = 2\ln r + \ln \text{constant}$

or $\tau r^2 = \text{constant (etc.).}$

Question 6

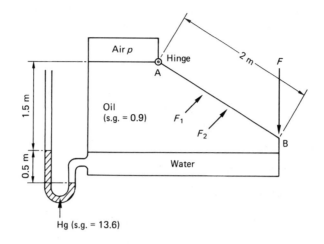

By manometric principles:

$$p + \rho_0 g h_0 + \rho_w g h_w = \rho_{Hg} g h_{Hg}$$

$$p = \rho_{Hg} g h_{Hg} - \rho_0 g h_0 - \rho_w g h_w$$

$$= 10^3 \frac{kg}{m^3} \times 9.81 \frac{m}{s^2} \ [(13.6 \times 0.5 \ m) - (0.9 \times 1.5 \ m) - 0.5 \ m] \left[\frac{N \ s^2}{kg \ m} \right]$$

$$= 48\,560 \ \frac{N}{m^2}$$

Owing to air pressure,

$$F_1 = 48\,560 \ \frac{N}{m^2} \times 2 \ m \times 1 \ m = 97\,120 \ N \text{ acting at 1 m from A.}$$

$$F_2 = \rho g \overline{h} A = 900 \ \frac{kg}{m^3} \times 9.81 \ \frac{m}{s^2} \times \frac{1.5}{3} \ m \times 2 \ m \times 1 \ m$$

$$= 8829 \ N \text{ acting at distance } x_0 \text{ from A,}$$

where $\quad x_0 = \overline{x} + \dfrac{k_g^2}{\overline{x}} = \overline{x} + \dfrac{I}{A\overline{x}} = 1 + \dfrac{bd^3/12}{A\overline{x}} = 1 + \tfrac{1}{3} = 1\tfrac{1}{3} \ m.$

Resultant force = 97 120 + 8829 = 105 949 N.

Point of application given by taking moments about hinge A:

$$105\,949x = (97\,120 \times 1 \ m) + (8829 \times 1\tfrac{1}{3} \ m) = 108\,892 \ N \ m$$

$$= \frac{108\,892}{105\,949} = 1.028 \ m.$$

Also, $\quad\quad\quad\quad\quad\quad\quad F \times 2 \cos 30° = 108\,892 \ N \ m,$

$$F = \frac{108\,892}{2 \cos 30°} = 62\,869 \ N.$$

Question 7

$$u = \sqrt{2gh_{\mathrm{m}}} = \sqrt{2gh\left(\frac{\rho_{\mathrm{m}}}{\rho_0} - 1\right)}$$

$$= \sqrt{2 \times 9.81 \ \frac{\mathrm{m}}{\mathrm{s^2}} \left(\frac{13.6}{0.9} - 1\right) \times 0.1 \ \mathrm{m}} = 5.26 \ \frac{\mathrm{m}}{\mathrm{s}} \, .$$

$$(Re) = \frac{\rho u_{\mathrm{c}} D}{\mu} = \frac{900 \ \frac{\mathrm{kg}}{\mathrm{m^3}} \times 5.26 \ \frac{\mathrm{m}}{\mathrm{s}} \times 0.25 \ \mathrm{m}}{0.03 \ \frac{\mathrm{N \ s}}{\mathrm{m^2}}} \left[\frac{\mathrm{N \ s^2}}{\mathrm{kg \ m}}\right] = 39\,450,$$

i.e. $(Re) > 2500$.

Hence turbulent flow.

$$u_{\mathrm{m}} = 0.8 \ u_{\mathrm{c}} = 0.8 \times 5.26 = 4.22 \ \frac{\mathrm{m}}{\mathrm{s}} \, .$$

$$\text{Mass flow rate} = \rho A u_{\mathrm{m}} = 900 \ \frac{\mathrm{kg}}{\mathrm{m^3}} \times \frac{\pi}{4} \times 0.25^2 \ \mathrm{m^2} \times 4.22 \ \frac{\mathrm{m}}{\mathrm{s}} \, ,$$

$$\dot{m} = 186.3 \ \frac{\mathrm{kg}}{\mathrm{s}} \, .$$

Question 8

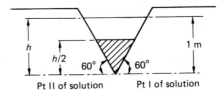

I: $\dot{V} = AC\sqrt{mi}$,

where $C = 60 \ \dfrac{\mathrm{m^{1/2}}}{\mathrm{s}}$, $i = \dfrac{1}{2500}$,

$$A = \tfrac{1}{2} \times 1 \times \frac{2}{\sqrt{3}} = \frac{1}{\sqrt{3}} \ \mathrm{m^2} = 0.577 \ \mathrm{m^2} \, ,$$

$$m = \frac{A}{P} = \frac{0.577\sqrt{3}}{4} = 0.25 \ \mathrm{m} .$$

$$\dot{V} = \sqrt{\frac{0.25 \ \mathrm{m}}{2500}} \times 60 \ \frac{\mathrm{m^{1/2}}}{\mathrm{s}} \times 0.577 \ \mathrm{m^2} = 0.346 \ \frac{\mathrm{m^3}}{\mathrm{s}} \, .$$

II: $$A = \frac{1}{2} h \frac{2h}{\sqrt{3}} - \frac{1}{4}\left(\frac{1}{2} h \frac{2h}{\sqrt{3}}\right) = \frac{3}{4} \frac{h^2}{\sqrt{3}} ,$$

$$P = \frac{2h}{\sqrt{3}} + \frac{h}{\sqrt{3}} = \sqrt{3}h ,$$

$$V = (AC\sqrt{mi})^2 = \frac{A^3}{P} C^2 i$$

$$0.346^2 = \left(\frac{3}{4} \frac{h^2}{\sqrt{3}}\right)^3 \times \frac{3600}{2500} \bigg/ \sqrt{3h} = \frac{3600}{2500} \frac{3h^5}{64} ,$$

180

giving
$$h = \sqrt[5]{\frac{0.346^2 \times 2500 \times 64}{3600 \times 3}} = 1.121 \text{ m},$$

and
$$\frac{h}{2} = 0.561 \text{ m}.$$

Recommended Reading

The market for textbooks in fluid mechanics is fully supplied with many volumes of adequate rigour and range of topics, and, as so often, it is very difficult to be dogmatic about the best possible purchase as far as a particular student is concerned. The choice is largely a subjective one, depending on the student's own particular likes in the manner of presentation and the drawing of diagrams.

Two useful books are *Mechanics of Fluids* by B. S. Massey (published by Van Nostrand Reinhold) and *Fluid Mechanics* by J. F. Douglas, J. M. Gasiorck and J. A. Swaffield (published by Pitmans).

Both of these are of adequate standard for normal degree courses in engineering fluid mechanics and only if advanced study is anticipated should the student look elsewhere; for a much wider coverage of gas dynamics the reader is referred to the very comprehensive text *The Dynamics and Thermodynamics of Compressible Fluid Flow* by Ascher H. Shapiro (published by Ronald).

I can only advise that the intending buyer should look carefully at any prospective purchase and see how the dimensional treatment is handled and if the text is in the SI system of units. One possible point to consider is that the older and more established a book is, the more likely (one hopes) that any numerical errors in the examples have been eliminated by continued correction and reissue. This factor must, however, be balanced on the other hand by the need for modern treatment in the matter of units and language which should accord with British Standards (if a British book is under investigation).

Index

Acoustic velocity 142
Analysis, dimensional 76
Approach velocity 44
Area, first moment of 12
Area, second moment of 14

Bearing, journal 37, 105
Bernoulli equation 33
Boundary layer xvi
Bourdon tube 12
British Standard 47
Buckingham theorem 101

Car jack 28
Centre of force 13, 15
Centroid 12
Centrifugal pump 34
Chezy constant 111
Clearance, radial 105
Coefficient of contraction 51
Coefficient of discharge 43
Coefficient of friction 78
Coefficient of velocity 43
Compressibility 1, 2
Continuity equation 32
Continuum xvi
Contraction, end 113
Contraction, sudden 60
Contracta, vena 43
Cooler, oil 89
Critical condition 144
Critical pressure ratio 147
Curved surface 14
Curved vane 74

Dam 25
Darcy 78, 123
Dashpot 129
Density 1
Differential manometer 11
Diffuser 141, 144
Dimensional analysis 76, 99
Dimensions xiii
Discharge coefficient 43
Dynamic force 76
Dynamic similarity 99, 101
Dynamic temperature 142

Eddies 44
Efficiency, nozzle 147
Efficiency, stagnation 150

Efficiency, transmission 82, 92
Elasticity, modulus of 1
Energy degradation 35
Energy equation xii, 33, 84
Enlargement, sudden 59
Enthalpy, stagnation 141
Enthalpy/entropy chart 155
Euler equation 32
Excrescences 85

Film, oil 130
First moment of area 12
Flow
  isentropic 41, 143
  laminar 122
  measurement 41
  non-uniform xv
  reverse 128
  sonic 143
  subsonic 143
  supersonic 143
  uniform xv
  unsteady xv
Flow nozzle 45
Fluids xii
  compressible 140
  ideal xiv
  real xiv
Force, centre of 13, 15
Francis formula 114
Free surface 110
Friction coefficient 78
Friction factor 85
Froude number 100

Gases 1
Gas constant, universal 143
Gauge, pressure 12, 19
Geometric similarity 101
Gradient, hydraulic 34, 79
Gravity, specific 1, 42
Gyration, radius of 14

Hagan-Poiseuille relation 125
Heads 34
  dynamic 41
  kinetic 41
  manometric 34
  potential 34
  pressure 34

total 34, 41
velocity 34
Heat transfer 136
Hydraulic car jack 28
Hydraulic gradient 34, 79
Hydrostatic equation 8, 9

Ideal fluid xiv, 3
Impulse 144
Inclined plane 130
Inertia force 76
Inlet port 105
Introduction xi
Inverted U-tube 11
Isentropic flow 143

Jet 57
horizontal 57
vertical 57
Journal bearing 37, 105

Laminar flow xiv, 122
Language xii
Leakage 132
Liquids 1

Mach number 76, 100, 143
Macroscopic approach xvi
Magnification, manometer 19
Manometers 8, 9
differential 11
inverted U-tube 11
pressure 10
U-tube 10, 41
Manning formula 143
Masonry bridge xii
Mass continuity xii
Maximum power 83
Measurement of flow 41
Microscopic approach xvi
Modelling 99
Model testing 100
Modulus, bulk 1
Molecular mass 143
Momentum equation xii, 55
Moody diagram 78, 85, 95
Moulding box 24
Mouthpiece 6

Newton's equation 55
Newtonian fluid 3
Nomenclature xiii
Non-uniform flow x, xv
Notch 104, 111
Nozzles 144
efficiency 147
flow 45
force 59

Number
Froude 100
Mach 76, 100, 142
Reynolds 76, 99

Oil cooler 89
Oil film 130
One-dimensional flow 31
Open channel flow 110
Orifice, sharp-edged 43

Parallel axis theorem 14
Parallel plates 122
Pascal's equation 8
Path line xv
Pelton wheel 92
Perimeter, wetted 77
Pipe flow 76
Pipes in parallel 81
Pipes in series 80
Pipes in series/parallel 81
Pitot tube 46
Pitotstatic tube 46
Plane, inclined 130
Plane surface 12
Port, inlet 105
Power, maximum 83
Power transmission 82
Pressure
centre of 13
hydrostatic 8
manometric 10
stagnation 142
static 34, 142
vapour 2
Pressure coefficient 76
Pressure force 76
Pressure gauge 12, 19
Pressure ratio, critical 147
Pressure wave 142
Propellor 107
Properties, of real fluids 1

Radial clearance 105
Radius of gyration 14
Real fluid, definition of xiv
Recommended reading 182
Reverse flow 128
Reynolds number 76
Rheological chart 4

Second moment of area 14
Sharp-edged orifice 43
Shear, viscous 122
Siphon 79
Sketch xii
Specific gravity 1, 42
Sonic flow 143

Stagnation point   46
Stagnation pressure   142
Stagnation properties   141
Stagnation temperature   142
Static pressure   142
Static temperature   142
Steady flow   xiv
Steady, non-uniform flow   xv
Steady, uniform flow   xv
Streamline   xv
Streamline flow   xiv
Streamtube   xvi, 56
Stress, shear   177
Subsonic flow   143
Sudden contraction   60
Sudden enlargement   59
Supersonic flow   143
Surface, curved   14
Surface, free   110
Surface tension   2

Temperature, dynamic   142
Temperature, stagnation   142
Temperature, static   142
Tension, surface   2
Thrust, hydrostatic   13
Torque   105, 134
Transfer, heat   136
Transmission efficiency   82, 92
Tube, pitot   46
Tube, pitotstatic   46

Turbulent flow   xiv
Two-dimensional flow   31

Uniform flow   xv
Universal gas constant   143
Unsteady flow   xv
Unsteady non-uniform flow   xv
Unsteady uniform flow   xv
Upstream velocity   44
U-tube manometer   10, 41

Vane, curved   74
Vapour pressure   2
Velocity   31
    acoustic   142
    approach   44
    sonic   76
    upstream   44
Velocity coefficient   43
Vena contracta   43
Venturi   41
Viscosity   2
    kinematic   3
Viscous flow   xvi
Viscous force   76
Viscous shear   122

Wave, pressure   142
Weir   111, 112
Wetted perimeter   77